ARCHAEOASTRONOMY

A study of The Alignment of Archaeological Monuments with the Sun's and Moon's Declinations
in Antiquity resulted in A Method to Estimate their Ages.

The Sun & Moon Events the Ancients Witnessed

(AN EXCERPT)

By Arturo Villamarín

ACKNOWLEDGEMENTS

I thank my brother Jorge for his assistance with the study and the travel arrangements in Colombia.
And, Ms. Lily Ickow for her suggestions on writing style.
And, Mr. Jairo Ivan Pachón of INSTOP Surveying Equipment Bogotá, Colombia, for his kind assistance with the provision of equipment for the measurements at El Infiernito.
And, Mr. Jens T. Satre for making the satellite calculator available on line, without which this work would have been not possible.
And, Google©earth Pro. For the original Panoramio Photos user database Layer (Now deleted), which served us as a detailed crowdsourced global survey of archaeological sites.

This is a selection of the complete results of research published under the same title.
Available at Amazon.com

ORCID: 0000-0002-5937-9110

ISBN- 9781723786198

ABSTRACT: A STUDY IN ARCHAEO ASTRONOMY

A method was developed to find the age of ancient monuments using the sun's declination in antiquity. This method showed the monument ages in the study are significantly older than those calculated via other methods. In this paper we present the results from the study of Teotihuacan in Mexico and El Infiernito in Colombia

Thousands of years in the past, the earth's axis due to the variation in its obliquity (nutation) and its precession, pointed to the celestial sphere in a different direction than it does today. These two factors made the sun rays impinge upon a site on earth at an angle different than it does today at any time during the year. Since the early days in human history it has been claimed, some archaeological monuments were designed in antiquity to align with the sun at certain times during the year. Those claims were and are mostly based on observations under *contemporary* astronomical conditions, not under the then prevailing conditions in the year they were designed in. Today those monuments have become misaligned since those earth axis' parameters have continued changing thru the ages; which makes those *contemporary* observations just approximations. Going back in time to find the original direction the earth's axis pointed to when the monuments were designed to align with the sun, allows us to calculate their ages.

Plotting the calculated sun's geographic positions during the solstice thru millennia with respect to a monument on a GIS map, generates a curve analog to an analemma, which we named a Super-Analemma. Extending a line on the GIS map at the azimuth of a monument's axis of symmetry to its intersection point with the Super-Analemma yields the year of the monument's original alignment with the sun; hence its age. Applying the method in various archaeological sites unexpected alignments of the sun were found in combination with those from the moon, in the years the method indicated the monuments were designed. These alignments validated some of those traditionally claimed singly for the sun or the moon, but with greater precision.

The Teotihuacan results show that the position of the pyramids in the layout, not only matches various astronomical phenomena but also show the arrangement reflects the planetary system. The connection between the events and the physical layout is punctuated by the Avenue of the Dead: Its length, its orientation and its displacement from the pyramids of the sun and Quetzalcoatl alignment.

The following sites are included in the complete study: Teotihuacán in México, El Infiernito in Colómbia, Mohenjo Daro in Pakistan, Machu Picchu in Peru, Giza in Egypt, Kalasasaya in Bolivia, Stonehenge in the UK and Tifariti in W. Sahara.

PREFACE

Since the day man landed on the moon in 1969 we gained the ability to *extract* ourselves from our history: contemplate the possibility that the authors of ancient monuments were not in our current linage. We are now able to hold the earth in our hands and look at it; see its weather, volcano eruptions and its position in space in relation to other celestial bodies. One picture which encapsulates this thought and which nearly all of humanity can recognize today is the photograph astronaut Michael Collins took of earth from Apollo 11; an earthrise over the moon's horizon as the Eagle descends to the surface of the moon.

"The only person who ever lived in the universe NOT in the picture..." NASA

Standing on the moon enabled us to detach our humanity from the geometric reality of these celestial objects. This study addresses the existence of ancient monuments from this perspective, in a detached manner the way the rover cruised the surface of the moon measuring everything without questioning how it came into being: an analytical approach. We hope to have met the ultimate objective; to address the eternal question "Why they exist?"

The study of ancient monuments, particularly those not associated with living quarters, present a distinct problem for anthropological studies. Unlike citadels, forts and cities, their purpose has not been obvious. The enigma they represent has been the Holy Grail of humanity since the beginning of time. For obvious reasons, their integration into the historical milieu of the societies which have flourished around them has been the accepted practice and approach to explain their existence. This has generated a myriad of inconclusive theories as to their purpose:

Life, death, rituals, religion, water, agriculture, fertility, procreation, etc. are some of the topics associated with those theories.

Arqueoastronomy since the early 1700's has attempted to formalize the study of these sites in a scientific context, finding in the monuments their connection with celestial bodies, following the approach envisioned by the early Greek and Roman Astronomers, such as Eratosthenes of Cyrene 276-194 BC -who measured the circumference of the earth by means of a clever experiment using an obelisk and a water well, or the erection of obelisks like the Montecitorio obelisk in Ara Pacis, which supposedly served as the gnomon of a sun dial to mark Emperor Augustus' birthday.

For this study we held the earth in our hand and found each structure and measured its geographic location and architectural properties. With the aid of easy to use, readily available on-line tools, we discovered what may be a good answer to the rationale for the existence of these enigmatic monuments. They are the record of astronomical events which took place in the past. Finding the date the events occurred in, will allow us to know the approximate age of those monuments. As markers, the monuments hold embedded in their conception and execution the astronomical and technological acumen of their authors. Teotihuacán, Kalasasaya, Moehnjo Daro, El Infiernito, Machu Picchu and other archaeological wonders, capture in their designs the sun and moon phenomena of the past. The Great Pyramids, Stonehenge and Teotihuacán also embed in their designs the makeup of the planetary system. This will be discussed for each in its corresponding chapter.

The method is based on the discovery of an easy way to determine the sun's position with respect to a point on earth in the year the inclination of the earth's axis pointed in a different direction than it does today with respect to the celestial sphere, due to its precession cycle (Hipparchus ~200BC). In that position the sun rays impinged upon a site on earth at an angle different than it does today; angle at which a monument could have been aligned to. Today those monuments have become misaligned since the earth's axis has continued changing directions thru the ages in its precession cycle. Going back in time to find the original spatial geometric conditions of earth; the direction its axis pointed to, the location in its orbit and its axis' tilt (Nutation from 21.5° to 24.5°) when the monument was designed to align with the sun, would allow us to find its age.

We used a Global Information System (GIS) Google©earth Pro and a satellite calculator to find the sun's position over millennia during the solstices relative to a point on earth. Plotting the position data generated a curve; an analemma analog, we called super-analemma. This is the precession's analemma. Extending a line at the azimuth set by the monument's axis of symmetry (or other major feature) to the intersection point with the superanalemma yields the year the monument was originally designed to align with the sun.

INTRODUCTION

The way the sun and moon illuminate archaeological monuments has been recognized since the early days of archeoastronomy as a salient characteristic at many sites. The light from the sun and moon accentuate one or several aspects of the monuments' design or layout during the year; an approximate geometric alignment occurs. The illumination, be it due to the sun's position at the site during the equinox or the culmination of the summer solstice or from the moon during its phases, has been given various interpretations by researchers and the public at large. The most popular and prevalent explanations for the purpose these monuments have are: the monuments are connected with agriculture, they served mystical purposes, for some others they exalt the renewal of life in the spring and still others give them religious meaning.

In the early 1700s W. Stockley a "Druid Revivalist" and J. Wood, a "Mystical Architect" were the first *scientists* to realize, places like Stonehenge had alignments with the sun and stars. John Wood declared "The ancients expressed all their knowledge in emblematic form in the lay-out of their temples; not only in the plans but in the astrological relationships between their sites", as reported by John Michell in the book titled: A little History of Astro-Archaeology[1, p. 12] This book provides a succinct history of the still prevailing struggle between traditional archaeologists, mystics and mathematically driven analysts, regarding their approach to the study of these monuments. We belong in the latter. The mathematical analysis as proposed early on by Sir J.N. Lockyer; the father of astro-archeology / archeoastronomy, is the vein in which this study was conducted.

In previous writings, we have described how many of these monuments may have had a more fundamental purpose: we showed, in some instances, they are astronomical markers. In this work we present the result of research which provides an improved site ageing methodology. The method shows the location and positioning (bearing) of these monuments were deliberately selected to reflect the sun and moon positions in relation to earth at the time each site's design was laid out. The monuments served and serve as astronomical historical markers. The designers incorporated in their designs the geometric measure of the celestial bodies' alignments with respect to earth, thus leaving a mathematical record of time and the sidereal position of the planet at different times during the year; echoing J. Wood's insight.

Dr. Sherry Towers of ASU published in her web site [2] a series of articles dealing with the mathematical analyses that are required to ascertain whether alignments exist for archaeological sites, primarily addressing those with stars with the physical features of any one of the well known sites. Rightly, she points out the movement of the celestial bodies through time need to be considered when assessing whether an alignment was built or designed into the monument.

> "Thus, for assessing whether or not an archaeological site has alignments to the rise/set of a particular star, all we need to know is the star's declination *at the year the site was built(our italics)*, and the latitude of the site".

In subsequent articles she provides the details of the math involved as well as the necessary refinements which are needed to account for terrain and atmospheric refraction.

A salient consequence of the well established hypothesis, by Sir J. Norman Lockyer: "Temples were aligned to their foundation date", is the reaffirmation of a generalized concept we had hinted at in the earlier writing: many if not all the archaeological sites appear were built thousands of years prior of their current dating. The discrepancy may be due to the inability of the current techniques to definitively tie the monuments to the societies who may have lived around them through millennia. Current dating techniques are based on various analytical methods: atomic decay of carboniferous matter, the stratigraphic analysis of soil deposits for the presence of various isotopes using radiometric aging and GPR -Ground Penetrating Radar; used to *see* inside monuments or below them without damage, among others. Carbon dating, the most widely used method, does not conclusively tie the structure with the civilization(s) that may have lived around them, it is plausible these civilizations may have modified or appropriated the monuments to suit their religious need or for other practical purposes. The association of the tests' results with the vestiges and with the monuments is typically done by considering the overall *context* for a particular site.

Logical hurdles exist to be surmounted in the analysis of geometric alignments. The cumulative error from the various methodologies used to measure the alignment of monuments with the sun, moon and stars, and the ultimate uncertainty of current dating methods, and not ignoring the probability of an alignment occurring by chance alone, as Dr. Towers [2] points out regarding analyses using the data, add difficulty in reaching definitive conclusions.

> "The problem with a site that has many associated site lines is that there is a high probability some of those site lines will line up with the rise/set azimuth of a star just by mere random chance. Alternatively, if we try matching site lines to the rise/set azimuths of many different stars (plus the Sun solstices and major and minor Moon standstills), we will get matches just by mere random chance. It is a very common mistake to dramatically underestimate these spurious match probabilities, and take as a given that whatever matches we observe must be evidence that the site was used as an observatory"

Sir Lockyer, Penrose and Wood and many others who have surveyed sites and found alignments, suffered in their conclusions, not only from the probability problem but the lack of technical capability to accurately calculate the sun and moon declinations *thousands of years in the past*. These considerations enlightened us in the development of a simplified model / approach to aid in obtaining a more accurate estimate of the age of major archaeological sites. We used Sun and Moon declination data obtained with modern technology. With this technology we hoped to attain truer *astronomical reverse engineering*. We found it relatively easy to get historical declination data for the sun and moon for thousands of years in the past utilizing readily available satellite calculators available on line [3]. These calculators are fairly accurate (1-2 arc minutes = .016666667deg.); artificial satellite positioning rely on these measurements for their exacting needs. For our purpose, we assume all the mathematics required; orbital parameters, perturbations, etc. are embedded in the calculator. Also, that the calculations remain valid when calculating thousands of years in the past. This last assumption is tenuous; the author of the mathematical equations used in it claims it is valid for only the 20th. and 21st. centuries[4]. A communication with the calculator's designer Mr. Jens T. Satre revealed it has not been tested for years earlier than stated. The results obtained in the analysis of several sites did not reveal obvious inconstancies in calculating saros cycles of the moon, the declinations for the sun and moon and eclipses; the results were consistent from site to site and with contemporary published data in the book Thousand year Cannon of Solar Eclipses by Fred Espenak, portions presented in the web site; http://www.eclipsewise.com. For our purpose, to prove the concept to a good approximation, it was deemed adequate.

This study was started by confirming known alignments of archaeological monuments, only with respect to the sun and moon. Also, a search was conducted for alignments at other archaeological sites where alignments may not have been recognized. We fail to envision a preponderant need the earthbound ancient engineers might have had in aligning monuments with the stars. For travelers from exoplanets, the stars would have served as astronomical landmarks which place the location of our solar system in space. The positions the stars have with respect to earth at specific times of the year and its use for ocean navigation is an ancient legacy concept, owing its origin, perhaps to those travelers. If, on the other hand, the knowledge was homegrown (an ancient technologically advanced civilization), from the time those first navigators took out to sea; the stars' *fixed* position in the sky served them well. From land, their usefulness would have been limited. We theorize that an understanding of the *inner* planetary system, earth, sun, moon, Venus and Mars, for either of these travelers, would have had a purposeful immediate use in their explorations, as this study has lead us to conclude.

Diverging from the established theoretical frame work seems to predispose some to reject new perspectives. We focus our studies only on that which can be measured and reproduced by the public at large, alibi the technological or informational limitations they may have. Our objective was to test our hypotheses in an easy to follow clear manner. Also hoped to demonstrate its value to most people and in particular to those who command the science and have 'high technology' at their disposal and can test our findings and reproduce them and minimize or eliminate the technical deficiencies which may be found in our results.

John Michell summarizes the controversy arising between traditional archaeologists and astronomers/surveyors thusly: "Archeology has this point in common with theology that it deals with questions which are not by their nature susceptible to first-hand proof. It is therefore even more inclined than most sciences to rely on dogma, on which account the boundary lines that limit its orthodoxy are more sharply defined" [1 p.79]

Having no ax to grind with either of these camps, we took a pragmatic approach for the study of archaeological structure alignments; not very unlike Hawkins [1, p66], but not as rigorous:

Apply geometry *(earth-measure) to the analysis of monuments found on earth. Find their location on the earth's sphere and their geometric and positional attributes and measure them! Determine if those physical measurements and or positions are related to any other monuments found anywhere on the sphere (solid geometry). Survey each site's layout and its geometric framework to determine or confirm the structure's alignments with the sun and moon. Find an explanation for the universality of these alignments and verify it; a rationale which would explain why global (unconnected?) primitive societies found it imperative to commit dearly scarce resources to projects of this nature, particularly to those of great magnitude.*

The underlying assumption (theory) to be tested is: the monuments are astronomical markers built disregarding the topographic features of their locations and the populations that may have lived around them. Today we place geodetic markers (NGS National geodetic Survey) having up to 1:10,000,000 positional accuracy; with the same philosophy but their locations are selected for geographic reasons.

In a way, this thinking goes a step beyond Alexander Thom's discovery that the extension of the alignments found in Stonehenge connected with other site's alignments. He found it to be "a vast system of astronomically placed stations extending far across the Wessex landscape..." [1, p.84] and the alignments found earlier by Sir Watkins; his lay line measurements [5]. In our publication; Nasca and Easter Island, An Ancient Global Plan Revealed [6, p.37], following these pioneers' vision, we brought into the methodology simple geometric definitions to the analysis of sites and lines:

Any line drawn on a sphere is an arc segment of a great circle. A line serves three purposes: sets the distance between two points, sets a boundary and or points in a direction.

With these arguments we discovered a picture of wider scope than these great pioneers had envisioned; we determined that Nasca and Easter Island are global maps to archaeological sites; back then astronomical markers. We showed how the direction of each line or Ahu following its great circle point to ancient locations around the globe. With this perspective, how we came about to aboard the astronomical alignment problem and the development of a method to find the age monuments were designed is discussed here. Detailed explanations are given with illustrations and examples for several major archeological sites, including maps, graphics and data, which appear within the text and in the Appendix.

The Hypothesis, its basis

The moon's orbit is at an angle of about 5^0 with respect to the ecliptic, therefore the earth and moon's orbital paths cross at two points; these points are referred to as *nodes*. When the moon is found near the nodes, it may *align* with the sun and the earth; these are the times when *eclipses* are likely to occur. Therefore, the two times per year the moon is found near either node are called *eclipse seasons,* these last about 40 days. The moon in its orbit is continually changing its position relative to the sun and with respect to earth. The position in its orbit determines the angle -declination angle- with respect to the earth's equator. Likewise, due to the earth's axis tilt angle, the sun appears to move up and down across the Equator changing its declination angle with respect to the earth's equatorial plane; its declination angle varies depending on the earth's position in its orbit. When both the sun and moon reach a maximum -or minimum- declination; the solstices (standstills for the moon) and at mid point; the equinoxes, each one of these determine *unique relative positions;* the three celestial bodies achieve a particular spatial arrangement at those times. From the observer's point of view; the sun and moon's elevation over the horizon in the sky (altitude) and the cardinal point direction (azimuth), determine a unique sun, earth and moon arrangement which lasts only short time periods, for any given location on earth. The same arrangement can be measured disregarding the observer's location by taking the measurements with respect to the center of the earth. These measurements are universal; that is an astronomical arrangement may be measured from any one location on earth, regardless of whether it can be observed from the selected location. These and other arrangements of the three bodies -with all the concomitant instantaneous parameters- re-occur only over very long periods of time, this makes each arrangement virtually unique today, as they were in the past and will be in the future. Solar or lunar eclipses or close approximation alignments are special events which we will show; the record of which was kept with these large *markers* we now call archaeological monuments, although in some instances, just stones oriented in a particular way record the events. The effect these astronomical arrangements have on the seasons and weather and the spectacular displays they bestow upon us, and which we surmised were represented with monuments, have consequently given us and others the impetus for their study. They have made many investigators and writers associate the monuments with agriculture and *local needs,* rituals and even the divine. These associations are secondary or incidental to the structure's existence but not the paramount reason for it. The positioning and design of archaeological monuments found around the globe is coherent; we will show they do not stand alone in each region where they are located. Affirming Sir Watkins' concept behind his Lay Lines [5] this research shows that the sun's position, in some instances, is triangulated from various points around the globe; that is, the bearing of monuments located in different continents is aligned to the same location the sun was found in at one time. The ultimate result, we find, is a monument revealing an astronomical alignment occurring elsewhere: a node point crossing causing a new moon and in another case an eclipse at a remote location not visible from the point of the structure's azimuth alignment with it.

From these and other results we conclude; the civilization(s) who engineered these monuments were highly advanced in astronomical sciences. The breadth of their reach was global; the method -the same at every location- used to capture the astronomical events reveals a cohesive plan. Perhaps they were interested in the weather of planet earth, the mechanics of which they left represented in these monuments. The alignments captured in their designs and/or layouts, show the effect the three bodies' position in space has on the seasons and the climate and also, on resulting phenomena such as solar or lunar eclipses and the phases of the moon.

This writing covers an incipient area of research based only on geometry which can be expanded with the appropriate computer analysis and modeling to fill-in the gaps and to tighten the error in measurement. These gaps are not only in the number of sites that were studied and to the limited extent the studies were carried out. Typical astronomical modeling will serve to confirm if the conjunction of the solar solstices and the moon's major and minor standstills or anyone of other specific arrangements, indeed do present unique convergences of events that can be geometrically associated with the monuments. Also, that these and other relationships were written in the physical measurements and positioning of the monuments; *in the language of mathematics* (Galileo). If confirmed, the events recorded in the monuments' designs, due to their uniqueness, can be used to identify the monuments' age.

The implied ability to predict eclipses, near misses, new and full moons and its major and minor standstills ensconced in these monuments, was perhaps left for the benefit of future generations. With this study we confirmed our earlier conclusion: the monuments are astronomical markers for the historical position of earth and the moon with respect to the sun in antiquity. Alexander Thom was, perhaps, the first to associate the moon standstill's positions with the geometric arrangement of stone circles; he envisioned these recorded the moon's rise and set times during those events:

"As I write this (1969), the Moon is coming through a major standstill. One cannot fail to be surprised to see it set and rise almost in the north. A fortnight later one is again surprised to see how far to the south are the rising and setting points, and how very low it is at transit."[7]

HYPOTHESIS DEVELOPMENT
Historical And Technical Background

On June 21, in the year 1800, during the summer solstice in the northern hemisphere, an astronomical phenomenon was observed that had not occurred in over sixteen years and would not repeat until two years later. The sun and moon were near their maximum latitudes simultaneously, the moon was approaching a major standstill which happened in 1802 and would not repeat after that until another eighteen years and seven months would have passed. In the summer of the year 1800 the sun and the moon travelled together in the sky for a complete day's cycle. This phenomenon *was apparently not common knowledge at that time*. This celestial phenomenon was discussed in the anecdotes of Dr. Parr appearing in The Gentleman's Magazine in October 1828, p.314: a conversation between a Mr. Bowles and Mr. Duke (Rev. E. Duke?) in reference to the stone circle of *Abury*:

"But, says Mr. Duke, intrenched in cycles and epicycles, "look here!-these two circles in another circle, represent the Sun and Moon travelling together!"-"The Sun and Moon never travel together two minutes", quoth Mr. Bowles"

We know they *do* travel together under certain astronomical conditions; when the moon crosses the Equator close to the vernal equinox, that year the moon achieves its maximum possible latitude and a *Major* or *Minor* lunar standstill occurs. These unique standstill types occur 18.6 years apart; under those conditions the moon's declination is near 0^0 at the Equator and so is the sun's, both, at the vernal and the autumnal equinoxes. That year two standstills, of the same type will occur, either Major or Minor. Major and Minor standstills alternate at 9.3 year intervals. During that year's solstices, the moon will also be at either one of these standstills. When these conditions are met and the moon is near a node the Right Ascension (RA) is 0^0 or 180^0, respectively; there is a high probability an eclipse will occur a few days before or after the day the equinox crossing takes place. With all these parameters, during the solar solstices, for about two years before and after a major or minor standstill, the sun and moon may be seen travelling together. Table 1 summarizes the data for the lunar events of 1800 and 1802, at Teotihuacán. Did the builders of the *Abury* stone circles have advanced astronomical knowledge, as judged by Mr. Duke?

Table 1

The above conditions are similar during the New Moon, there is a good likelihood for the three bodies to fall in line, the sun and moon appearing to be close to each other in line with the earth, but the moon's standstills won't be at maximum latitude. Then, the moon is illuminated from the back, at near 0^0 from the line of sight from a point on earth. Even if it is skewed by a few degrees, the sun's brightness prevents naked-eye sight. During this phenomenon the two bodies maintain a narrow angular distance measured as a difference in azimuth (elongation).

New moons are particularly important, in this study because they cannot be seen, unless an eclipse occurs. This study will show that the designers of archaeological sites knew with precision the exact location of the moon at those times. When the moon is close to the orbital nodal points, at the intersection point of the earth and moon orbits the moon aligns with the sun then, the three bodies fall *in-line*; this is known as sun-moon-earth syzygy; an alignment of three or more celestial objects. If the moon is in between the sun and earth, a solar eclipse occurs. If the earth is in between the two, a lunar eclipse occurs. A precise alignment in the three spatial directions yields total eclipses of the sun or moon. Small misalignments yield partial eclipses. In solar eclipses the moon's shadow covers a relatively small area on earth. The moon's shadow is small so it paints a narrow shadow band of about 250 km. in width as it moves from west to east. The moon orbits in the same direction as the earth, however it *lags* behind, so, to an observer on earth it appears to rise in the east and move west. The earth's shadow is much larger than the moon's size in the sky, so a total lunar eclipse can be seen over most of the side of the earth facing it. From an observer's (topocentric) point of view, when the azimuth difference (t-Right Ascension; defined as the longitude of the moon measured starting from the sun's position at the vernal equinox; at the point the sun crosses the Equator from south to north) between the sun and the moon approaches $0°$ or $180°$, if an eclipse occurs, it can be viewed from the site, if the alignment falls within certain parameters. Some type of eclipse can occur at a longitude around $9.92°$ about a node; ascending or descending, i.e. moving above or below the ecliptic. From a geometric perspective the evaluation of eclipses is *geocentric;* the center of the earth is the reference point for the location of bodies in space along the Cartesian coordinates: x,y,z. The geocentric evaluation of phenomena is absolute, independent from the observer's point of view and without atmospheric diffraction, topographic conditions or parallax. In this study we use the geocentric frame of reference, except where explicitly noted. The eclipse ranges used are tighter; without parallax, topography or atmospheric diffraction.

"For any type of solar eclipse to take place, the Moon's (ecliptic) latitude must not exceed 1°28' N or S. When new moon occurs within 18°31' longitude of either of the nodes, a partial solar eclipse is possible if the Moon's latitude is small enough. Within 15°21' of a node, a solar eclipse of some kind must occur. A total or annular eclipse occurs when the Moon's latitude is within 58' N or S and a new moon falls within 9°55' longitude of a node'. The actual value around the node ranges from 15.39° to 18.59°. For any type of lunar eclipse to take place the Moon's latitude must not exceed 56' N or S and a full moon must occur within 11°38' of the nodal line. For a total lunar eclipse, the Moon's latitude must not exceed 26' N or S and the full moon must fall within 3°45' of the nodal line." [8]

In earlier work [9] we confirmed there is coherence in the placement and alignment of *some* monuments among themselves; the geographical positioning of three or more sites generated great circles on earth. We concluded that for this to be true their alignments had to be geometric, with the center of the earth as the reference point: geocentric. Most of the data presented here is therefore calculated as such; topocentric data is also presented and in some instances compared. This preliminary conclusion needed further proof as to the purpose for the alignments of those *geodetic markers.*

In the summer of the year 1800 in Teotihuacán, México, we found the sun and the moon kept a distance between 7^0 and a maximum of 21.7^0 during the day; this translates to an apparent distance of about 1200 miles, between their positions on earth at their zeniths. The close distance was maintained throughout the day; which indicated the moon was approaching a node and or a standstill. The angular distance between the two celestial bodies is dependent on their positioning with respect to earth and the simultaneous movement of moon and earth around their orbits. The arrangement of the earth and moon with respect to the sun in space, at anyone instant, can be described in terms of several parameters; among them, the precession of the earth's axis, the earth's axis tilt -the sun's declination-, the time of year, the moon's position (phase) during its *lunation* cycle and the moon's declination with respect to earth. In one instance, the apparent distance between the two bodies diminishes as the sun's declination reaches its maximum or minimum latitudes and the moon also reaches its corresponding maximum or minimum declination *within the same time frame*. As the sun reaches its maximum declination (solstice) it remains *fixed* long enough at a *maximum declination* and the moon can reach the same declination on its way to its own standstill, its maximum declination about 5.1^0 higher in latitude. If this occurs near a node -the moon's ecliptic latitude approaches 0^0- a solar eclipse may take place. The solar max and min happen at the culmination of the solstices; one time each per year; summer and winter.

The Moon's orbital mechanics involved are significantly more complex than it may appear at a glance. A *Synodic month* is the time the moon takes around its orbit from new moon to new moon (29d 12h 44m 03s). The moon's orbit around the earth is not fixed; it varies constantly in various respects; therefore all the values quoted are mean values. Its orbit is not circular and its degree of eccentricity varies constantly, which causes its distance to earth to vary by approximately 40k kilometers with an average distance between them of 384,400km. Its orbit inclination varies, as well; its mean inclination angle is 5.145^0 with respect to the ecliptic with a maximum of 5.3^0. Due to this orbit tilt the moon crosses the ecliptic twice in a month. The moon changes from upper to lower culmination in less than four weeks 27.212 days -a *Draconic month*; defined as the time the moon takes for one revolution from its ascending node across the ecliptic (27d 5h 5m 36s). Due to the orbit's eccentricity, as explained, every other New Moon the moon is at its closest distance to earth -Perigee- and the intervening month at its farthest -Apogee. The time the moon takes to reach from Perigee to Perigee is defined as an *Anomalistic month* (27d 13h 18m 33s). The longitude of the ascending node is synchronized with the sun's longitude, starting with 0^0 at the vernal equinox. The duration of a Draconic month is longest when their longitudes are either 0^0 or 180^0 and shortest when the angle is 90^0 or 270^0. The orbit's inclination changes in equal manner with respect to these angular differences. At angle differences of 0^0 and 180^0 it nears its maximum of 5.3^0 and a minimum of 5^0 at 90^0 and 270^0; the result is, *the solar or lunar eclipses occur when the lunar orbital inclination is near its maximum*. The combined effect of all the variances in each of the three lunar month periods: Synodic, Anomalistic and Draconic is known as a harmonic which repeats every 18.6 years; which is equal to the time it takes the orbit ascending node to reach the same location; in a retrograde precession of the orbit. This is known as the Saros Cycle. Therefore, its cyclical nature is due to the moon's orbital inclination to the ecliptic and the precession of its orbit. The moon's two different maximum declination points are known as the major and minor lunar standstills, i.e. when the moon is above the ecliptic and the earth's axis tilts in its direction and when the tilt is away from it, respectively [10].

Finding Sun and Moon Alignments in an Astronomical Hay Stack

In antiquity, before the Babylonians 1600BCE or the astronomers; Ptolemy100AD and Galileo 1564AD, the prediction of eclipses and the reason for them, was totally unknown (Thales is credited of having predicted a total solar eclipse in 585 BCE). However their scientific foundation and the geometry of the alignments were not defined until about 1609 by Johannes Kepler. The possibility of those alignments having been recorded in the designs or the layout of monuments, which were built thousands of years prior to these early astronomers, is minimal.

Although arguably impossible for the ancient designers of pyramids and stone circles to have been able to predict the sun and moon alignments, or for an eclipse to have occurred at anyone location and its geometry recorded; nevertheless we find the occurrence of either kind of phenomena, were actually captured in the design and alignment of monuments. Many have argued against this claim based on some of the given reasons and others. It is also argued that alignments may occur by chance alone, Dr. Sherry Towers of ASU, stated in reference to rising star's alignments "we will get matches just by mere random chance"[2].

However, the mathematics which predict the precise moment in time when the three celestial bodies come into an alignment are real. When we find the phenomena occurred and the geometry of the three bodies' position in space, at the time it occurred in, is also found in the geometry of the monument erected at the site where the phenomenon happened, the geometric pairing is a fact that cannot be ascribed to chance. Predicting all the mathematical variables coming together and occurring at a specific time and place for eclipses to happen, is a mathematical reality; eclipses can be predicted years into the future or calculated in the past. A skill the ancient architectural designers had to have had to plan the building of such monuments. In the study we will show not one but multiple phenomena alignments at a single site and at multiple major sites, the world over. For simpler alignments of the sun or moon by themselves with monuments, the combined evidence of these on a global scale is preponderant, even if a good number of these are random. The case in point are the well documented *fairly good* alignments the sun and moon display with monuments during the solstices and equinoxes, regardless of the latitudes the monuments may be found in; from Brodgar at latitude of 59°N to Tiwanaku at a latitude of 16.5°S

To find sun or moon alignments during the course of this investigation, the specific monument was selected in sites where multiple choices are available. *The year the site may have been designed-* as determined by the method; the subject of this writing, was used. Each site was tested for sun-moon position combinations as outlined in the protocol in conjunction with the basic observation times: sun/moon rise, maximum elevation during transit and set time. The breadth of this study was limited to a scant set of well known sites and only a limited number of test variables were applied. The study was initiated in Teotihuacán, México. The technical basis for the study, as explained earlier, was to test the use of the geometric astronomical positioning of the three bodies -without the effects of parallax, topography or atmospheric diffraction- as an approach to define the age of archaeological sites. The alignment of monuments with the sun and moon during astronomical phenomena, evaluated under these conditions; the result of their spatial geometry, we expected would prove our hypothesis.

To determine the exact time and location the three celestial bodies have at any one moment *manually* is a slow process, subject to the mathematics involved in the evaluation of the planetary mechanics. The ancient engineers achieved it. Given the arguments laid out, it is remarkable that the positions represented in arrangements of the three celestial bodies, such as a solstice and standstill, appear to have been accurately recorded in antiquity at several locations around the globe and frozen in time in the monuments' designs. The preeminent site where this became obvious to us was Teotihuacán. At the site, the Pyramids' layouts reflect the astronomical geometry of the sun and moon positions at given times during the day when both bodies are near their maximum latitudes simultaneously. The designer's objective, we surmise, was to leave a permanent legacy of our solar system astronomy, which reflects in its rendition all their concomitant technological knowledge. This knowledge is implied in the reckoning of the phenomena and its occurrence having been captured in the alignment of the monuments. The designer's technical acumen is demonstrated in the sheer magnitude of each rendition. Ultimately, the designers may have meant to show the effect the variations in the astronomical arrangements have in creating the earth's weather; a legacy for those who might in the future inhabit or visit this planet. They built *gigantic obvious models* intrinsically beautiful to represent the astronomy behind the phenomena. The development and realization of such global *instrument* presents an ample motive which justifies their existence and the magnitude of these enterprises; dwarfing any heretofore explanation.

Historical Background. The Alignment of Ancient Monuments With the Sun

The azimuths of the sun and moon have changed over time with respect to a given site's *layout and or with the monuments' structural feature(s).* In the study we measured the change in these angles at dawn, noon and dusk. In previous studies we had noticed that the sun's angle of incidence at those times quite often impinges on archaeological monuments either perpendicular or parallel to their sides or geometric axis (as observed in the 1800's by the researchers Wood and Lockyer [1, p.20]) or the sun and moon azimuths are in line with elements of the overall layout. For example; in the Temple of Kalasasaya in Tihuanaku, Bolivia, at sunrise during the summer solstice in the southern hemisphere, the sun simultaneously aligns closely with the Ponce Stele and the Gates of the Sun and the Moon, we reported earlier [9, p.152]. As a consequence of the new method, the subject of this paper, the incidence angles will be fine tuned and a complete analysis will be provided in the chapter dedicated to this monument. During that study it did not go unnoticed that several archaeological sites have dedicated sun and moon monuments; such as the one just mentioned and Teotihuacán; which is arguably one of the most grandiose. The simultaneous presence, at these sites, of monuments dedicated to each heavenly body suggested the astronomical relationship that exists between the two, could be embedded in their designs, i.e.; New and Full Moons and eclipses.

Teotihuacán is not aligned north as most sites are, making it rather unique; consequently, it was selected as the first test site. Incidentally, the results of this study took us beyond the development of a way to find the age of monuments. We discovered Teotihuacán's layout, represents the Sun, Earth, Moon system; an explanation for this conclusion is provided after we discuss the basis for it. We will describe how the positions of both the sun and the moon, when they are found near their maximum latitudes in the north, an arrangement found in the protocol; during the summer solstice in this instance align with features of this archaeological site.

In this part of the study, the sun and moon alignment data shows the effect *time*, thru millennia, may have had on the alignments. Their positions, as measured *today* (the year 1800), have changed with respect to their theoretical original alignment at the time the monuments were designed. The study shows that the geometry of the sun-earth-moon relationship has changed over millennia, at any time of the year. This will become apparent as we describe the geometry of *today*. In the analysis we use the words 'almost' and 'nearly' quite often; simply because when we describe, for example, 'the sun impinges perpendicularly' it is more often than not, from a fraction to one or two degrees off perpendicularity. In an earlier study, we had hypothesized, that this slightly skewed geometry could reflect the passage of time and believed that a time regression analysis based on the geometry of a structure and the position of the sun in the past could tell its age, particularly if one could tighten the source of error in the measurements. John Michell relates, Sir Lockyer, had come to a similar hypothesis when he analyzed Plutarch's description of Hathor: [1, p.20]

> "Hathor, according to Plutarch, is Isis, and at Dendera in the temple of Isis is the inscription, 'Isis shines into her temple in New Year's Day and she mingles her light with that of the father Ra on the horizon.' Ra is the sun and Isis, otherwise Sothis, is the star Sirius. Lockyer calculated that Sirius rose in line with the extended axis of the temple of Isis in about700BC, agreeing with the archaeologists' date for the foundation of the temple and that it rose at the same time as the sun , thus proving the inscription to be a true record of an astronomical event"

We theorized the purpose behind aligning archaeological sites and their monuments with astronomical arrangements of the Sun, Moon and earth could have been to preserve the knowledge of those arrangements and other celestial phenomena derived from them, such as eclipses. The monuments would *cast in stone* the position of the three celestial bodies particularly at times which could have been significant; such as the solar solstice and lunar standstill. For our work the conjunction of the solstice and lunar standstill positions became the starting point for the search of alignments. We, also, used the eclipse seasons when the moon is near either of the nodes and its ecliptic longitude difference approaches 0^0 or 180^0 with respect to the sun.

As noted earlier, we believe there is error in measuring alignments of the sun and or moon or stars with ancient monuments' or their site's geometries based on the *contemporary* relative position of the three celestial bodies, not as they stood in the past. There is, also, *error* implied in the topocentric measurements. By definition topocentric measurements are localized to the observation point; it assumes the monument's designs were based on local observations. It does not consider the planetary geometry; the geocentric coordinate system. Our argument, not only eliminates the topocentric *error*, but, the conventional (egocentric) *error* in the common assumption, held up to now and virtually embraced by every one: the monuments are the product of our ancestors; the local societies who may have lived around them. Also, that if these societies did build them, they had no knowledge of the planetary system; although, the concept of their similarity to the planetary system -at Stonehenge- was suggested early on by John Wood and reaffirmed by Rev. E. Duke in his Druidical Temples of Wiltshire in 1846 [1, p.17]. The work presented here in combination with our previously published work [6], reaffirms the global design concept and the graphic representation of the planetary system, not only at Stonehenge but at Giza and Teotihuacán in their layouts, as we will propose.

What follows is the work that shows that applying the method the relative positions of the sun, moon and earth serve to explain the geographical positioning and the azimuth of the various archaeological monuments analyzed in the study.

The evaluation of alignments of the sun and moon positions with monuments via the method using the solar declinations calculated in the past and following the protocol as will be described in this writing provides a mathematical approach to estimate the age of archaeological monuments.

THE AGE OF ARCHAEOLOGICAL MONUMENTS

The results of the analysis of Teotihuacán's pyramids, show that when the sun and moon are near their maximum declinations, the earth sun and moon geometric relationship, causes the sun and moon positions to match, to a close approximation, the pyramids ' geometric lines and their layout at various times during the day. This analysis provided astronomical support data to the idea; the ancient monuments were designed to reflect the sun and the moon positions with respect to earth at specific times.

Alignments similar to the ones discussed have been recognized to exist at various locations globally and were studied by early researchers since the early 18th. Century [1, p.12]; their analysis began with basic empirical observations. These researchers were attracted, initially, to the concepts found in existing ancient Roman writings, legends, myths and popular folklore. Their approach evolved culminating in actual analytical measurements of the monuments' alignments with the sun, the moon and the stars' positions at various times. The alignment of the sun with an archaeological structure at sunrise during the solstice is the most widely recognized; it is a phenomenon which appears *was designed* into each monument's alignment and is still observed today at many sites around the world, regardless of the latitude they are found in; a fact which, in its totality, speaks of a *global design*. From a global perspective, our analysis confirms the *geometric relationship* between the sun and the moon's position is embedded in the monuments' designs in several archaeological sites. For some of these sites the alignments of the moon in relation to a monument appear to be emphasized over those of the sun. Not surprising since the moon alters the sun-earth astronomical relationship in many ways, of which eclipses are the most obvious.

In earlier work [14], we reported the results of an analysis of the alignment of the Intihuatana stone in Machu Picchu, Perú, with respect to the sun's azimuth at sunrise and sunset. That study was based on the physical measurements of the stone made by Rolf Müller in the 1920's. There we argued the presumed alignment of the stone with the sun's azimuth at sunrise, could be used as a tool which would provide a way to estimate the age of the site, in a manner similar to what Rev. William Stuckley had tried at Stonehenge, except he used the magnetic field variations to calculate its age at about 460BC [1, p.9]. If the intent of the designers of monuments was to record the sun's azimuth at sunrise during the summer solstice, at anyone location where these alignments are found; Stonehenge, Machu Picchu, etc. then, a misalignment of a few degrees could indicate the time it took for the structure -a fixed point on earth- to become misaligned. Such a calculation combined with the effect of the axis' tilt variation over time could provide an insight as to the structure's age, we conjectured. The mathematical analysis would entail a calculation akin to the calculation for the precession of the earth's axis. Precession causes a shift in the axis' direction in relation to the positions of the constellations in the celestial sphere. The earth's axis is currently pointing past the star Polaris by 44'; measuring the change in rotation angle over time served to calculate the nearly 25,700 years it takes for a full precession cycle to complete. During this long cycle the earth's axis orientation changes with respect to the sun, as well. Since the beginning of recorded history about 9,000 years starting with the early settlements, the earth's axis has completed less than half of a full precession cycle.

If the earth had a fixed position in front of the sun -no orbit- precession by itself would account for a change in the position of the sun with respect to a location on earth; a *secondary declination* up and down across the earth's Equator. Each quarter of a precession cycle the axis changes the direction it points to continuously, with respect to the sun. At one point the northern hemisphere is exposed toward the sun at maximum earth latitude; the earth's axis points towards the sun. A quarter cycle later the axis points perpendicular to the direction of the sun (tangential to the orbit); the sun is in line with the Equator. After this point the sun's latitude increases towards the southern hemisphere; the earth's axis now points towards the sun in the south until it reaches minimum latitude. For the second half of the cycle a reversal occurs and the sun moves back to the equator and past it until it reaches maximum latitude in the north, thus completing the 25,700 year cycle. Because this cycle is so lengthy, its effect is almost negligible in short time periods of a few centuries, i.e. the direction the earth axis points to, can be considered as being almost *fixed*. The planetary mechanics involved give rise to the *Precession of the Equinoxes (General Precession);* which move westward over the centuries.

In a manner similar as described above, but now considering the earth's axis tilt *fixed in one direction,* as the earth completes a cycle in its orbit around the sun, causes the sunrays to change angle - declination- with respect to earth giving rise to the seasons. The direction of the earth's axis tilt appears *fixed* in a year's time, which causes the sun's rays to illuminate perpendicularly at different latitudes within a declination angle interval of $\pm 23.5^0$ during the year. *The earth's position in its orbit* around the sun is the main reason for the change in the seasons in a year's time. This *primary* solar angle declination is the one we are all most familiar with as it affects our everyday lives. The *secondary* declination, described earlier, becomes useful when we start looking thousands of years into the past, as Sir Lockyer did, as he attempted to find the age of Stonehenge [1, p.23]. For a basic overview of earth mechanics good videos are available on YouTube [17]. Due to the precession cycle the seasons drift very slowly away from their given month names; the solstices and equinoxes happen in different months. Because calendars were designed to account for the *drift*, winter will always be in December in the northern hemisphere. To find the sun's declination in the past we used a sun and moon calculator [3]. The results were plotted in Google©earth to find the relationship between the sun and moon's geometric positioning to the site's geometry.

The above concepts are the basis on which the archaeological monument ageing method was developed. The following sections describe its development and application at Teotihuacán, followed by other major sites.

Analyzing Teotihuacán's Layout Design

In addition to the sun and the moon alignments with the pyramids, the Avenida de los Muertos and the site's layout, described earlier, we found other alignments, *whether these alignments were visible from the site or not at the time they occurred in* (geometric vs. visual). The fact that alignments were found, when the sun and moon were below the horizon, we judged could be significant; our *global geocentric design* hypothesis could be tentatively supported if it was possible to determine the alignments were not due to regularly reoccurring orbital cycling.

To find a precise alignment of the sun's azimuth with the Avenida de los Muertos' azimuth of 15.15^0, we used the calculator to map the position of the sun during the solstice, every 15 minutes for twenty four hours. Earlier, by extrapolating back in time we had found, in the 1800's, the 15.15^0, alignment occurs over three minutes before noon. A similar exercise had revealed that the moon's position during the solstice on June 21, 1800 was at latitude 25.76^0 longitude -96.99^0 and from this location it aligned with the Avenue of the Dead, at 15.17^0. At this latitude the moon was not at standstill; which begged the question: what year, day and time did it reached a Major Standstill and aligned with the Avenue? At this latitude this year, the moon was lower in latitude than the theoretical mean maximum at standstill of $\pm 28.5^0$. It is theoretical because there is a normal variation due to, nutation, an earth axis tilt variation, orbital precession and other factors such as orbit anomalies. Not forgetting *time* which becomes a major factor when the moon or sun's positions are calculated thousands of years in the past.

We selected the year 1800 serendipitously based on Dr. Parr's account; it lead us to a year the sun and moon travelled together. We plotted -in Google© earth- the sun and the moon's positions and their azimuth in relation to the Sun and Moon pyramids' coordinates, by running a line from each of their apexes to the geographic coordinates of the point which marked each of the sun and moon locations. We started at noon during the summer solstice on June 21, 1800, every 100 years into the past and five hundred years into the future, including the current year 2015. It was found that in the year 1800 *at noon* (12:00) the sun reached maximum latitude of 23.462^0N longitude 98.48^0W, whereas the sun's position at the same time, in the year 2015 reached latitude 23.437^0N and longitude 98.303^0W; the latter longitude is closer to the longitude required for alignment with the Avenue of the dead of 97.73^0W. In the year 1800 the sun reached this longitude minutes earlier 11:56:38 at the given latitude. In the year 1000 the sun's latitude was lower at 23.366^0N and in the year 2500 will be lower still at 23.292^0N and longitude 97.738^0W; unexpectedly, we found in this position the sun will align perfectly with the Avenue's azimuth of 15.15^0.

These results were indicative that this particular sun alignment in the year 2,500 with the Avenue could have also occurred in the past, at a different latitude and time. One of those previous occasions, we conjectured, would have been the year the design of Teotihuacán was laid out. We could have, painstakingly, plotted positions for the sun and moon into the past and tweak the hours, minutes and seconds until each point location for the sun and moon would have fallen in line with the Avenue: a daunting task (a program could be written to do the iterations), but there was no telling when to stop. To find the *stop year*, using the satellite calculator and Google© earth, we plotted* the sun's location every thousand years in the past and into the future. We discovered what soon became obvious; the sun's position on June 21 for every year, through thousands of years past and future, outlined what we came to call a **Super-Analemma** curve for the sun's position -the *secondary declination* shown on Fig.8. The date (June 21st·12noon) due to the axis's precession moves along the curve through millennia; rather, *the solstices and equinoxes drift through the months; the point of Aries moves in retrograde*. For the year Teotihuacán was designed the culmination of the summer solstice occurred on *September 4th.* will be shown later. The current Gregorian calendar is the last revision of a *calendar* which has had various versions since Neolithic times. Calendars, by convention, are used to keep track of time while fixing the 'drift' of the seasons and time. Leap year corrects for the drift in time in the short span of our recorded history. In early history the Egyptians used the star Thuban (Alpha Draconis), to indicate north since the earth's axis appeared to point in its direction. Due to precession, the earth's axis now points past Polaris or North Star by 44', it has moved away from Thuban to past the North Star and continues moving towards Vega and will reach it in about 13,000 years. The *circular* motion (Nutation makes the circle fluted like a coffee filter) axial precession or axis wobble is due to the axis tilt and the attractions of the sun and moon and to a lesser extent the pull the other planets exert on earth. The slow circular wobble affects the declination of the sun with respect to earth, changing the times the seasons occur in, as the earth orbits around the sun ~25,700 times (years) for one full precession cycle.

The sun's position plot through millennia -the super-analemma- is analogous to the sun's position analemma one can plot in one year at a given location. The one significant difference is that this super-analemma is not a closed loop; after about 30,000 years it advances creating another loop.

*No references found for numeric or graphic data for the sun's declination over millennia, as described in the text. The use of a graphic method to calculate the age of archaeological monuments graphically appears to be unprecedented. Several on-line calculators, including the Navy's, were tested to find the solar declination data. Of those, only the one quoted [3] is capable of accepting any year.

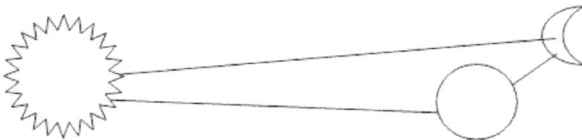

METHOD: APPLICATION in TEOTIHUACÁN, MĒXICO

In previous research we had discovered some of the Nasca Lines are arc segments of great circles drawn on earth [9]. Some of the circles generated from some of these *arc segments* follow lineups of global markers; archaeological sites. This was mathematically tested; plotting the geographic global coordinates, for each of those *marker* sites, on a 3D Cartesian coordinate grid resulted in circles. A typical plot may be seen in Figure 1

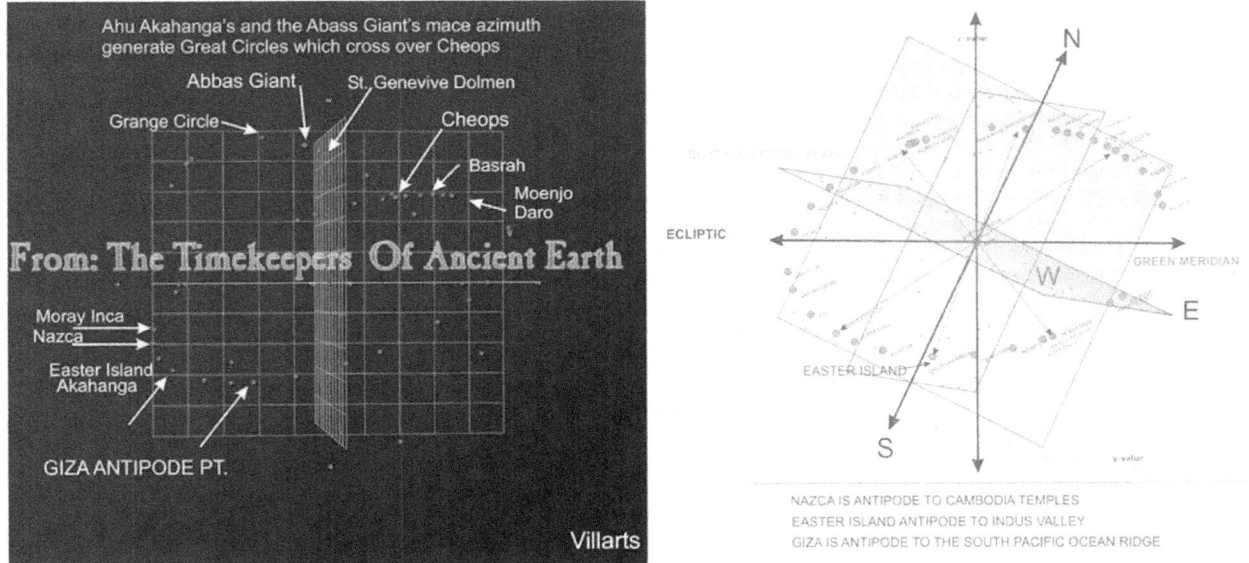

Figure 1 - A Great Circle in Ahu Akahanga's direction connects archaeological sites.

The Avenue of the Dead is, also, an arc segment of a great circle. It has an arc length of 2,157mts, it is 18.6 times smaller than the earth's circumference of 40,030mts. A lunar Saros Cycle is 18.6 years long. The moon's Saros Cycle appears to be codified in the length of the Avenue of the Dead, which starts at the Pyramid of the Moon's apex, ending at the southwest edge of the Quetzalcoátl temple. Drawing a great circle using the Avenue's azimuth of 15.45^0 measured by Millon, 1973:53, crosses even closer over Mehrgarh and Mohenjo Daro, than one drawn using our measurement of 15.15^0. The accuracy of this value will affect the calculation of the site's age. We tested a range of values from 14.4912^0 to 17.395^0 which yielded age dates within a 250 year range and within this range we evaluated nine different years and up to four months around a test point in the protocol. Using the 15.42^0 value, we traced a revised great circle -the *extended* Avenue- around the earth over the pyramid of the moon to find its intersection points with the

21

super-analemma. Another great circle perpendicular to the Avenue and the Pyramid of the Sun's axis was also drawn. The result is shown in the graphic, Figure 3.

The Avenue of The Dead Great Circle's Intersection with The Super-Analemma Yields The Site's Age

We hypothesized that the Avenue's azimuth was dictated by the positions of the sun and the moon at sometime in the past, when the positions of the sun and moon could have had similar geometry to today. The two intersection points of the Avenue (the Great Circle) with the super-analemma would provide the years and times that determined the Avenue's alignment with the sun. Once the year was determined by the intersection point, the moon's position was found using the calculator. The upper intersection occurs in the year 2500AD. The other intersection would have occurred at the time the original layout was designed; when the earth's axis pointed in a different direction due to its precession cycle and the sun's maximum declination was reached at a different time of the year. The lower intersection of the Great Circle -the extended Avenue line- with the super-analemma is the 'stop' point we would have arrived at had we continued with the iterations we mentioned earlier. However, ascertaining the validity of the intersection point-date still requires several iterations to test around the point-date, particularly if none of the expected sun-moon position combinations are found in the resulting year, i.e. the sun and moon do not display any phenomena such as eclipses or major or minor standstills. Also, as noted earlier, there is no agreement as to what the actual azimuth of the Avenue is, so a range of values was tested.   The data plot in Google©earth, Figure 2, shows the intersection occurred on June 21, -7,210, which would date Teotihuacán at 9,225 years old. This is the year the sun and the moon aligned in a unique way which may have prompted the designers to capture in the site's design and the alignment of the Avenue of the Dead. The details leading to this interpretation of the results is discussed later. We, also, tested the great circle perpendicular to the Avenue and the sun pyramid to find the intersection points with the super-analemma. A more recent date was found, but there was no supporting data to justify it. However, this great circle traces the maximum and minimum latitudes achieved by the sun that year. This circle sets the *Latitude* of the Pyramid of the Sun, and the entire complex, on the great circle generated by the Avenue.   The Equation of Time (the analemma) plot for the year - 7,210 for the 21st. day of every month, for twelve months, shows that the sun's maximum latitude (summer solstice) for that year occurred at noon on September 4 at latitude 24.637^{0}N, ** whereas the minimum (winter solstice) occurred on February at latitude 24.334^{0}S. These latitude values are close in value to the max and min latitudes reached by the great circle perpendicular to the Avenue. The next day September 5th.at noon the moon achieved its maximum latitude of 28.337^{0}N -a lunar standstill. As before, the sun and moon travelled together at a close distance throughout that day. At sunrise on this day; 5:14:00 and moonrise at 5:19:00 the sun and moon azimuths were: 63.4175^0 and 59.995^0 respectively.

**Current modeling indicates that the earth's "(axial)tilt reached a maximum of 24.2^0 about 9,500 years ago, oscillating $\pm 1.3^0$ from an average value of 23.3^0 [8]. This closely coincides with the sun's declination data the method yielded as the year Teotihuacán was designed in.

Early in the morning of September 21, at 02:00 the moon's position was at 25.364^0S 110.8^0W, it had an elevation of 42.86^0 and an azimuth of 194.719^0; this is the azimuth of the Avenue of the Dead in the southwest direction. The moon first appeared from the southwest aligning with the Avenue of the Dead facing northeast in the direction of the 'would be' pyramid of the Moon. At noon, the same day, the sun aligned with the Avenue of the Dead from the northeast, with an altitude of 86.186^0 and was located at latitude 23.356^0N, longitude 97.749^0W, the sun had an azimuth of 15.15^0 northeast. Three months later on December 24 at 05:15 the moon aligned with the Pyramid of the Moon and the Avenue of the Dead from the northeast with an azimuth of 15.171^0 during its *standstill at latitude 28.774 ⁰N*** and it had a geometric elevation of 80.41^0.

*** The maximum latitude the moon could have achieved the year -7210 would be $\pm 29.9^0$ as shown in reference [8].

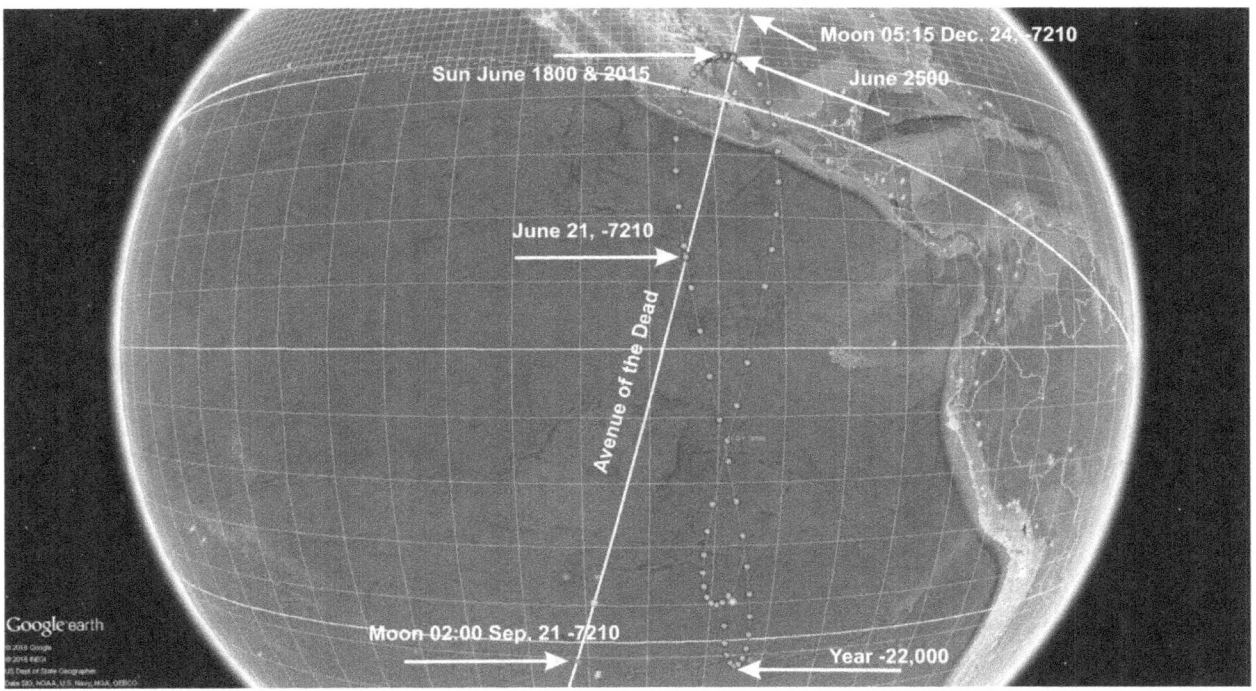

Figure 2 -The Super Analemma

The sunrise that day occurred at 06:11 with an azimuth of 99.73^0. In that direction the sun, also, aligned with the Lamanai Mask Temple in Belize, in the Yucatan Peninsula. This temple is itself aligned at about 100^0. About an hour later the sun had an azimuth of 106.21^0 and it was located at 9.515^0S 28.868^0W. In this position and at this angle it impinged exactly perpendicular upon the southeastern faces of both pyramids; the sun was on the eastern side of the great circle perpendicular to the Avenue. Aligned in this direction are several other archaeological sites; the

Lost City of Tayrona in Colombia, the Caracol in Belize, the Pomoná and La Venta pyramids in the state of Tabasco, México; the monuments at each of these locations face the sun at this angle. The great circles -in white- cross over the Moon and Sun Pyramids are shown In Figure 9 below. The curve is the analemma for the year -7,210. The inset shows Teotihuacán's layout.

The search for other astronomical phenomena in the same year yielded a total eclipse of the moon on January 26th. At midnight, the eclipse reached its maximum occultation over Teotihuacán, Fig.4. and Table 2. The moon was located at latitude 20.317^{0}N and longitude 97.228^{0}W with an azimuth of 67.0^{0}; Teotihuacán's coordinates are 19.69^{0}N and 98.84^{0}W and the azimuth of the NE diagonal across the Moon Pyramid is 65.6^{0}

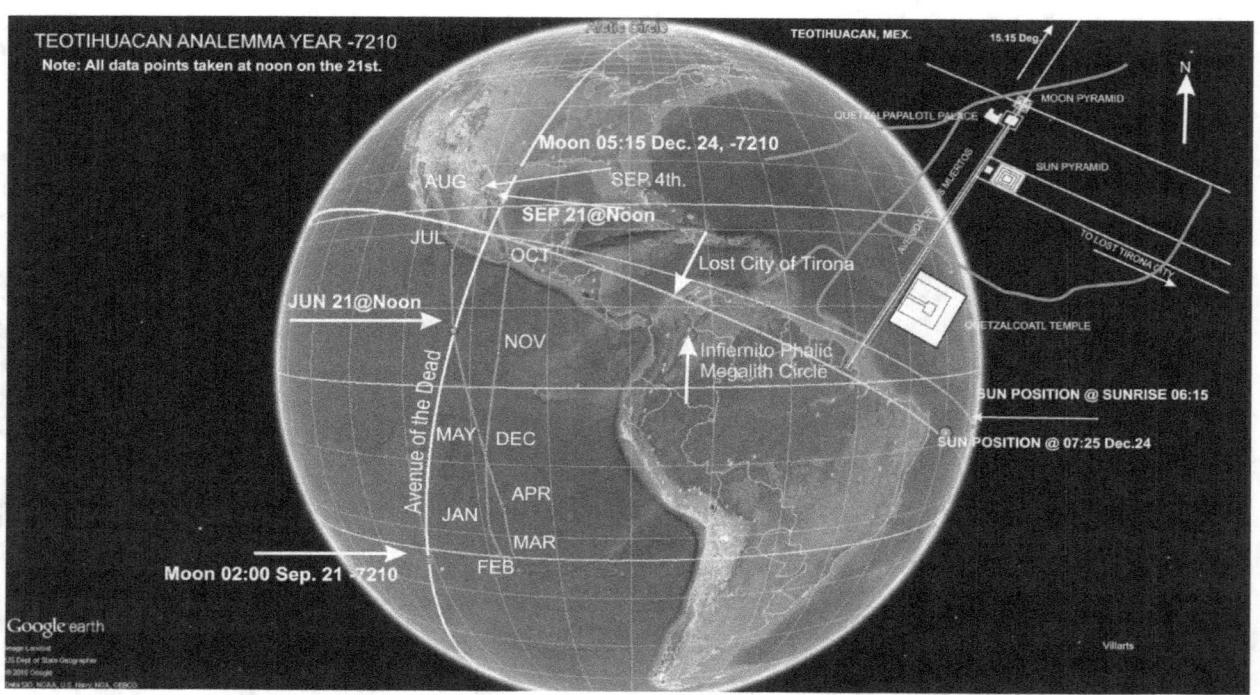

Figure 3 - Teotihuacán: Analemma For The Year 7210BCE And The Avenue's Great Circles

Seconds later the moon aligned at this azimuth and was physically located nine miles south of the Pyramids of El Tajín at 180^{0}; this site is located 175km (108 miles) to the northeast of Teotihuacán. The *Ball Court* at the center of El Tajín citadel is aligned NS. The umbral eclipse maximum occultation lasted for one hour and forty two minutes. The total eclipse from first to last contact lasted about four hours and twenty eight minutes. The results, thus far, indicate the year 7210 BCE is the likely date the site was designed. From the overall design perspective this doesn't explain the layout or the positions of Moon/Sun Pyramids in it: The moon pyramid heads up the arrangement, the Sun pyramid is east of the Avenue; the two monuments are not on the same axis of symmetry.

To find the alignments quoted -shown in the previous graphics, monthly sun and moon position data were collected for the period January to October. Table 2 shows the data segment of interest; the eclipse of January 26, the standstill and new moon of September 5th 7,210 BCE.

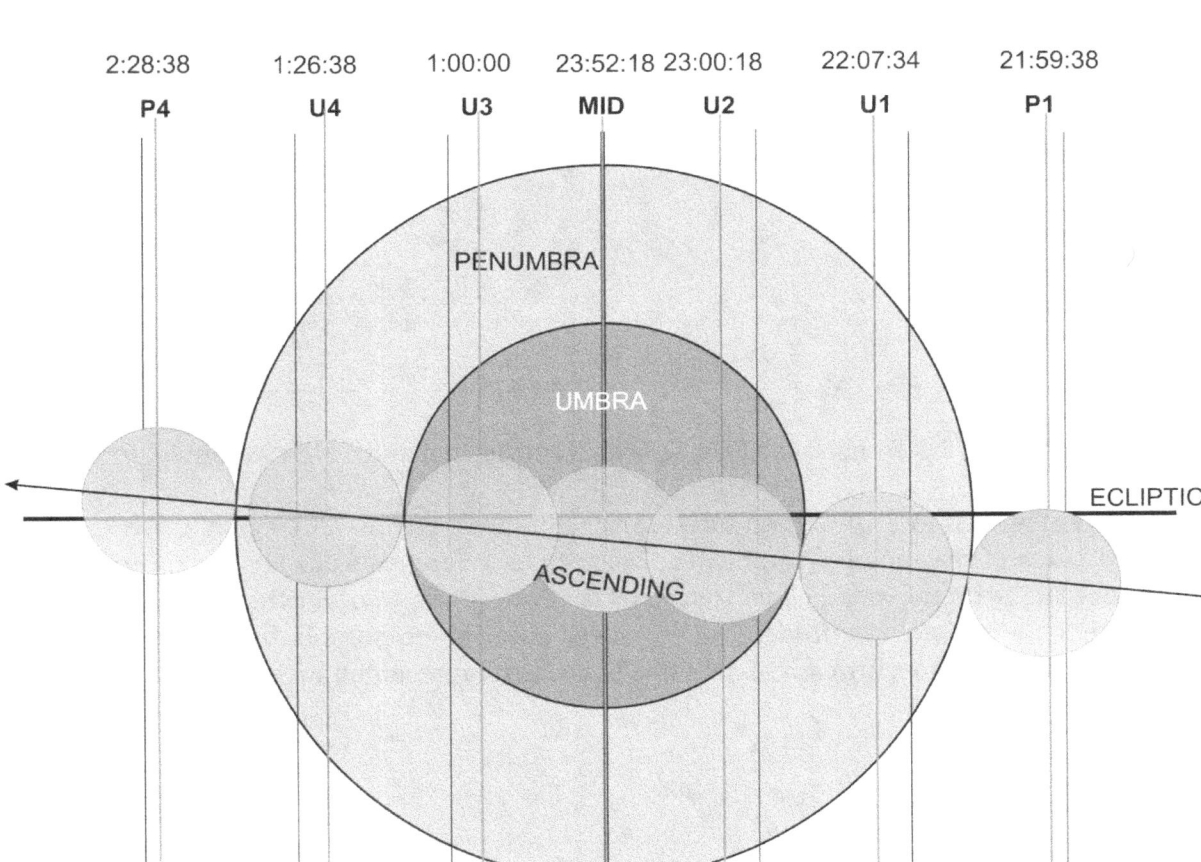

TEOTIHUACAN TOTAL LUNAR ECLIPSE JAN26, 7210 BCE.

Figure 4. Total Lunar Eclipse Over Teotihuacán, Jan 26 year 7210BCE

This data set, besides providing supportive evidence to prove the age finding method, it also supports the hypothesis regarding the pyramid's alignment with sun and moon being embedded in their designs. Also, it yielded a graphic pattern for the sun and moon positions which would be expected when the two bodies achieve narrow angular positions in the sky: *The New Moon in this case.*

When both the sun and moon positions are plotted in Google© earth at around noon time, the moon lags about 20 minutes behind the sun arriving at the point where it aligns with the Avenue of the dead. The moon moves at about 1,022km/hr. and the sun moves at 1,668km/hr. in (apparent) opposite directions. There is a point in time when the sun crosses its path with the moon's path, i.e. their azimuths are the same relative to a given point on earth or geocentrically with respect to the center of the earth. At that point if their elevations are close enough and are near a node, a solar eclipse could take place. Since the earth and moon orbits are skewed about 5.14^0 a solar eclipse doesn't happen every time they cross. For our purpose the question was; did an eclipse or New Moon occur on September 5, -7,210 and where did the sun and moon cross with respect to Teotihuacán? Or, at what time their paths superimposed; had the same *ecliptic longitude*.

Table 2 Total Lunar Eclipse Over Teotihuacán and a New Moon

Was this event related to Teotihuacán's design?

The coordinates of the moon pyramid were selected as reference to locate the positions of the moon, since the pyramid is dedicated to it. For the New Moon to occur the sun and moon need to have the same *ecliptic longitude*. The crossover point is measurable from a given reference point; topocentric or geocentrically. Could this have been represented with monuments?

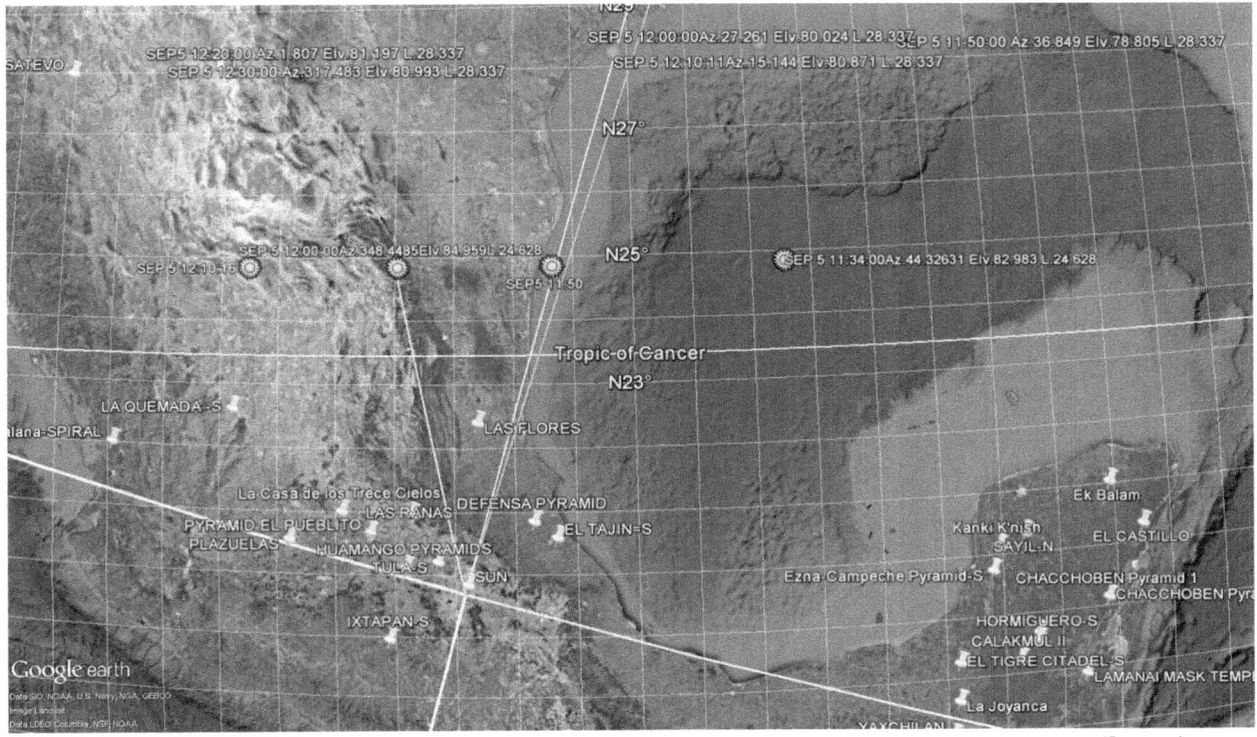

Figure 5. New Moon September 5, 7210 BCE Sun and Moon Locations Over Teotihuacán

Calculating the two bodies' positions every hour, it was determined that on September 5th - 7,210 the sun and moon reached the same azimuth 17.396⁰(diff.0.00095⁰), at 00:50:20 Teotihuacán's time on the other side of the earth, *just west of Mohenjo Daro!* At that moment, the sun and moon positions were points on a great circle with an azimuth which crosses over the pyramid of the moon. This azimuth angle was perhaps the reason behind the alignment of the *group of monuments aligned at 17⁰* mentioned earlier; Aveni, reported are found in Mesoamerica. About seven minutes earlier the sun had an azimuth of 15.1356⁰ and the moon 15.428⁰ an angular difference of *0.29238⁰and elevation difference of 2.889⁰. Nearing the New Moon, the sun and the moon aligned; they had the same azimuth as that of the Avenue of the Dead...(Moon!)* The New Moon doesn't shine, it is dead, and hence it could be the origin of the Avenue's name. As measured over Teotihuacán the alignment neared an eclipse initiating on the other side of the world, exactly over Mohenjo Daro; but there the sun and moon were north and south of the city and had a declination difference of over 3⁰, it was approaching a New Moon. The New Moon was reached at 2:48:55, the moon and the sun travelled together from that point on, which could not be seen at Teotihuacán until sunrise at 5:14:00. The sun rose first followed by the moon. At 11:49:41 the sun was over the Avenue of the Dead at Teotihuacán; latitude 24.628⁰N longitude 97.375⁰W and azimuth 15.13838⁰ and the moon reached a similar alignment of 15.55⁰ over the Avenue at 12:09:52. The data shows the site's axis of symmetry was designed to point to the beginning of a New Moon; therefore, the name should become *Avenue of the Dead Moon, i.e. killed by the sun* or just obscured by it in the sky. Many conjectures are found in the literature as to the origin of the name; this new interpretation should correct the others.

Incidentally, the name Mohenjo Daro was translated by Gregory L. Possehl to mean *Mound of the Dead Men* [35], perhaps it was Dead Moon?

Figure 6. The New Moon over Mohenjo Daro Determined Teotihuacán's Layout

The astronomical events just described confirm the date Teotihuacán was designed in: it supports the theory and the Method. Furthermore, our analysis shows that the geometry of the sun and moon pyramids' layout closely represents the geometry of the exact position of the sun and moon at Mohenjo Daro. To confirm this observation, we drew a right triangle over the Teotihuacán pyramids; the hypotenuse connects their apices, the adjacent is perpendicular to the Avenue of the Dead and the opposite side runs perpendicular from there to the Moon Pyramid. Similarly the positions of the sun and moon were connected with a line and each in turn with Mohenjo Daro's position; this resulted in a right triangle, as well. These two triangles appeared to be *geometrically similar*. Referring to the graphic, Figure 6; the sides of the dotted triangles were measured and the ratio of their sides for each were calculated. The ratios were found to be: Teotihuacán triangle; 1 : 3.09 : 3.25, Mohenjo Daro triangle; 1 : 2.89 : 3.1, these triangles share the same line (a side); the Avenue of the Dead. The hypotenuse at Mohenjo Daro becomes the opposite side to the Sun-Moon and Sun-Avenue angle. This occurs because throughout the day the sun and moon increased the time difference they had at Mohenjo Daro's longitude slowly, starting at about nine minutes difference represented by the 15.15^0 to 17.4^0 angles, ending with a difference of abut twenty minutes.

This analysis provides a justification for the Pyramid of the Sun's location being offset from the layout's axis and the distance and angle between the pyramids and the moon's prominent position .i.e. the moon's position higher in latitude. The question remains; why didn't the designers use the 17^0 angle? We can only speculate that the measurements were carried out at Mohenjo Daro, which could have been easily performed in the valley. The actual sun-moon crossover position - ecliptic longitude difference equal to 0^0- occurred over the Sulaiman Range the mountainous region shown in the map. In Teotihuacán's location the angle difference between the alignments of about two degrees; geometrically, it would have not worked out. The great circle perpendicular to the avenue over the sun pyramid, which records the solstices in the year 7210 BCE would have been off by that difference in azimuth. Instead, the 17^0 were recorded in the azimuth of other monuments. Fig.5

The Major Standstill of 7207 BCE

The moon's high latitude at standstill (28.774^0N) on September 21st. -7210, signaled it was close to its 18.6 year cycle and would reach a major standstill soon thereafter. Three years later on October 26th -7,207. a major lunar standstill took place. On this date at 8:00 the moon was aligned with the Avenue of the Dead with an azimuth of 15.075^0 had a geocentric declination of 29.729^0N and longitude 95.684^0W. See Table 3.

In years when major or minor standstills occur, the moon's position at the equator matches the solar equinoxes; vernal and autumnal. Due to this there is a high probability for alignment; eclipses may occur if the moon is near one of the nodes; this may happen twice during the year.

In the year -7,207 two eclipses took place. On May 21 a partial umbral eclipse, which could not be seen in Teotihuacán, was seen at Mohenjo Daro. Fig.7; the moon was over the Indian Ocean 184^0 due southwest of the stupa. At the start of the umbral phase the moon aligned at 15.15^0 with respect to Teotihuacán; it aligned with the Avenue of the dead, in the same manner it had happened three years earlier. This time, however, the sun was opposite on the other side of the earth over the Pacific Ocean, with an azimuth of 195.675^0 also *simultaneously aligning* with the Avenue from the southwest.

On June 4th. the day before the Vernal Equinox a partial solar eclipse of magnitude .5 took place. The eclipse lasted for 50 minutes from first to last contact. Fig.8

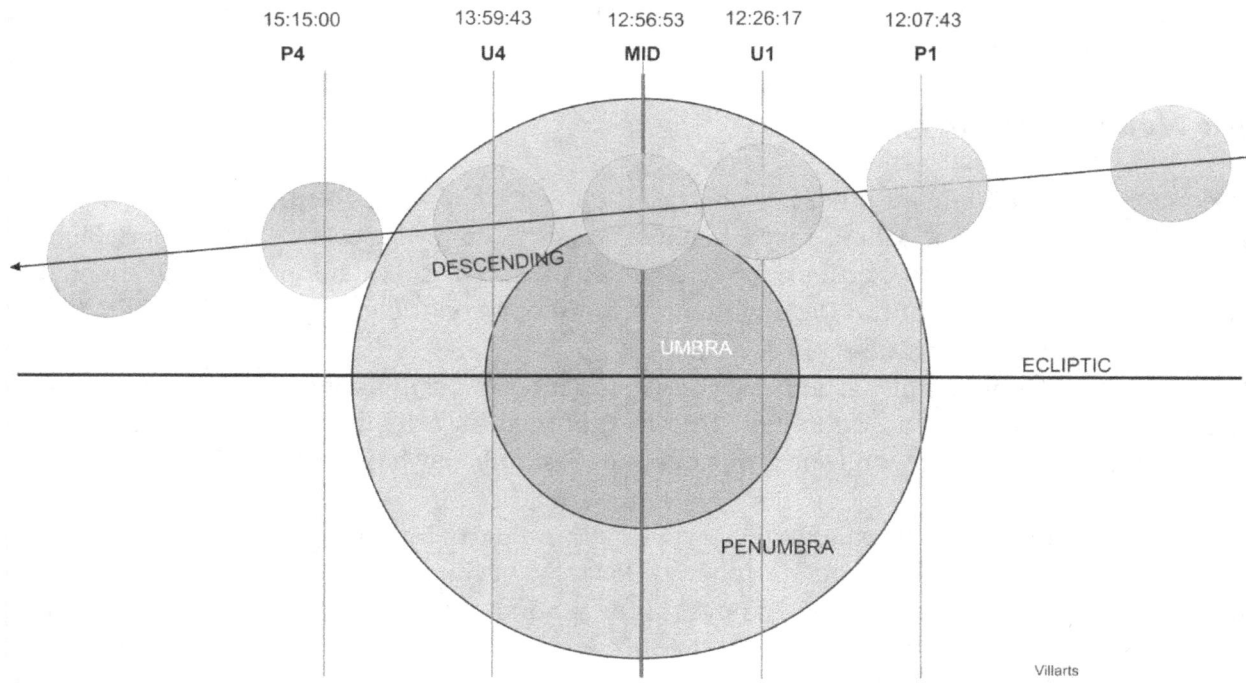

Figure 7 - Partial Umbral Eclipse at Mohenjo Daro, Not Seen in Teotihuacán

The second Major Standstill that year occurred on October 26th., the summer solstice had occurred on September 3rd.. The standstill was slightly higher than the first one, which was a southern standstill at 29.709°S, this *major standstill* was in the north at latitude 29.729°N and *the moment it was achieved, its azimuth with respect to Teotihuacán was 15.075°, the azimuth of the Avenue of the Dead and had a maximum altitude of 79.4°.* This standstill answered the question we had posed early on: what year, day and time did the moon reach a Major Standstill and aligned with the Avenue of the Dead?

Table 3 shows the eclipses and the Major Standstill data. The data shows there is a close connection between Teotihuacán's design and the positions of the sun and moon at various times when the two bodies reach their maximum latitude resulting in repeated alignments with their astronomical positions resulting in important phenomena. The results not only demonstrate the basis for the relative positions in the layout of the two pyramids and the site's alignment azimuth at 15.15°, but serve to support the viability of the *Method To Estimate The Age of Archaeological Monuments.*

The combined results also serve as a foundation for an analysis of the overall layout. Early on we stated; the overall layout "represents our planetary system" Admittedly, the resulting *Age* via the method and the graphic analysis we are about to present are contrarian to most everything that has been written -or we are aware of- regarding Teotihuacán's age and the cultures which may have lived around the site. The overall results and the 'Advanced Civilization' theory offer a basis for an *alternate or parallel* explanation for the layout given above.

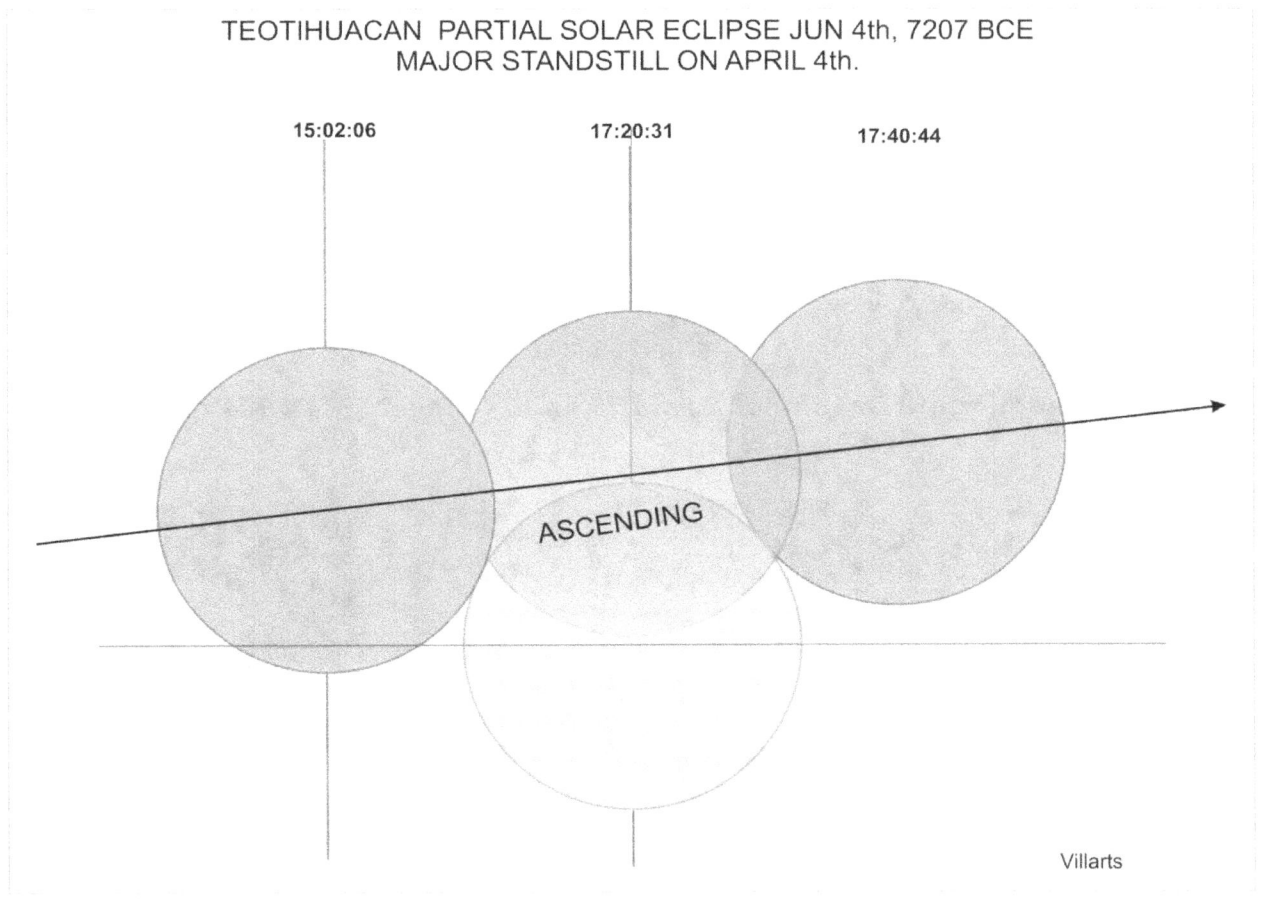

Figure 8 - Partial Solar Eclipse on June 4th. After The Major Standstill of April 4th.

6:35:22 19.6924 -98.844

YEAR									MOON DATA												
-7207	DATE	HOUR	AZIMUTH	S LONG	ELEVATIC	RA	DECLINA	longitude	HOUR	AZIMUTH	ECL LAT	RA	ELEVATIC	DECLINA	Longitude	E. LONG	e.long dif	DEC. DIFF	Az. DIFF.	Elv. Diff	RA Dif.
PENUMBRA P1		21:59:38	261.11	236.579	-48.083	234.02	-20.362	125.907	21:59:38	82.391	-0.24	52.363	48.887	19.712	-55.059	54.909	181.67	40.074	178.719	96.97	181.657
UMBRA U1		22:07:34	263.697	236.625	-63.95	234.068	-20.373	108.907	22:07:34	84.808	-0.175	53.068	64.429	19.953	-71.635	55.612	181.013	40.326	178.889	128.379	181
UMBRA U2		23:00:18	264.624	236.66	-76.301	234.104	-20.382	95.725	23:00:18	85.677	-0.125	53.617	76.526	20.137	-84.516	56.157	180.503	40.519	178.947	152.827	180.487
MID TOTAL ECLIPSE	26-Jan	23:52:18	244.396	236.695	-88.369	234.141	-20.39	82.726	23:52:18	67.002	-0.075	54.159	88.354	20.317	-97.228	56.695	180	40.707	177.394	176.723	179.982
UMBRAU3	27-Jan	1:00:00	95.4431	236.741	-75.563	234.188	-20.401	65.801	1:00:00	276.077	-0.01	54.868	75.957	20.55	-113.78	57.395	179.346	40.951	180.634	151.52	179.32
UMBRA U4		1:26:38	95.7665	236.759	-69.324	234.207	-20.406	59.143	1:26:38	276.382	0.015	55.147	69.868	20.64	-120.28	57.671	179.088	41.046	180.616	139.192	179.06
PENUMBRA P4		2:28:38	97.776	236.801	-54.832	234.25	-20.416	43.644	2:28:38	278.316	0.075	55.799	55.752	20.85	135.406	58.313	178.488	41.266	180.54	110.584	178.451
MAJOR STANDSTILL	4-Apr	18:10:44	105.121	301.16	-22.732	303.633	-20.9	8.774	18:10:44	123.444	-5.072	269.474	2.791	-29.709	-24.361	269.541	31.6190	8.8090	18.3233	25.5230	34.1590
VERNAL EQUINOX	4-Jun	13:00:00	221.755	359.522	64.114	359.566	-0.199	-115.75	13:00:00	223.564	0.436	358.534	63.051	-0.192	-117.04	358.588	0.9340	0.0070	1.8086	1.0630	1.0320
FIRST CONTACT		15:02:06	212.112	359.51	66.857	359.554	-0.204	-110.9	15:02:06	214.521	0.421	358.391	65.825	-0.275	-112.26	358.422	1.0880	0.0710	2.4091	1.0320	1.1630
		16:50:44	247.349	359.582	46.746	359.62	-0.174	-138.07	16:50:44	248.112	0.506	359.197	45.924	0.189	-139.04	-0.652	0.2340	0.3630	0.7634	0.8220	0.4230
MID SOLAR ECLIPSE		17:20:31	252.176	359.602	40.165	359.638	-0.166	-145.52	17:20:31	252.869	0.53	359.418	39.478	0.316	-146.37	-0.398	0.0000	0.4820	0.6934	0.6870	0.2200
		17:40:44	254.933	359.616	35.606	359.651	-0.16	-150.57	17:40:44	255.606	0.546	359.567	35.018	0.402	-151.34	-0.226	0.1580	0.5620	0.6726	0.5880	0.0840
LAST CONTACT		19:22:06	265.453	359.683	12.108	359.712	-0.132	-175.92	19:22:06	266.168	0.625	0.318	12.116	0.834	-176.16	0.637	0.9540	0.9660	0.7153	0.0080	359.3940
SUMMER SOLSTICE	3-Sep	7:00:00	70.6576	89.582	22.442	89.54	24.637	-25.219	7:00:00	62.778	4.376	108.437	6.054	27.928	5.347	106.275	16.6930	3.2910	7.8796	16.3880	18.8970
MAJOR STANDSTILL	26-Oct	8:00:36	84.4881	143.97	29.631	146.53	14.194	-35.66	8:00:36	15.075	5.131	86.506	79.412	29.729	95.684	86.908	57.0620	15.5350	69.4131	49.7810	60.0240
MOON@NODE@EO)	29-Nov	13:40:00	232.272	178.839	60.402	178.945	0.484	121.839	13:40:00	237.344	-1.032	175.444	57.486	0.952	125.752	175.462	3.3770	0.4680	5.0717	2.9160	3.5010
NEW MOON	29-Nov	19:50:07	279.835	179.101	-24.509	179.183	-0.859	144.864	19:50:07	278.801	-1.357	178.618	-26.377	-0.859	143.473	179.101	0.0000	23.6500	1.0338	1.8680	0.5650
AUTUMNAL EQUINOX	30-Nov	17:15:00	265.489	179.999	12.393	179.999	3.9E-05	-175.67	17:15:00	254.931	-2.434	189.699	17.823	-7.088	-166.83	191.747	11.7480	7.0880	10.5584	5.4300	9.7000

TABLE 1 - TEOTIHUACAN YEAR 7207 BCE ECLIPSES AND MAJOR LUNAR STANDSTILLS

Table 3. Lunar Major Standstills in the Year 7207BCE

A parallel explanation: Graphic Information and Geometry Is All We Need

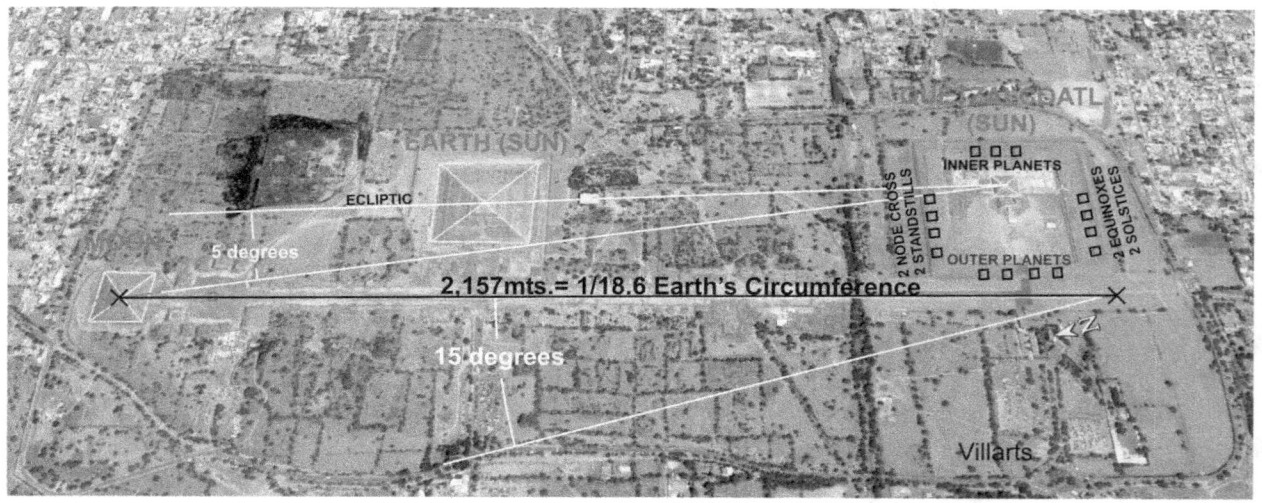

Figure 9 - Teotihuacán's Layout depicts the Planetary System

We start with a description of the entire layout. There are three main pyramids: from the north to south, The Moon, The Sun and The Temple to Quetzalcoatl. The "Sun" and Moon are *relatively* close to each other, while the Quetzalcoatl pyramid is significantly far apart. If we rename the Sun pyramid to 'Earth' Pyramid and the Quetzalcoatl Pyramid to Sun Pyramid; the layout of the three pyramids is that of the Moon moving towards an eclipse or new moon; this is supported as described earlier. The information in the geometric arrangement, also, justifies it. The relative distances between the pyramids support this view, plus the fact that both the 'earth' and the 'sun' are on the same axis (the ecliptic). The moon's orbit is tilted vs. the ecliptic;

therefore, graphically it sits to the side from the Sun/Earth axis, at an angle of about 5° which is its orbit's minimum value [2]. See Fig.9.

Now, focusing on the Quetzalcoatl pyramid we find that the total area encompassing the Temple is numerically and visually significantly larger than the Sun pyramid's base: 154,557 sq.mts vs. 54,680 sq.mts. This layout makes a relative graphical representation of size and distance. Facing the Temple, far away from the pyramid there are four small flat-top pyramids and another three similar in shape and size, in back of it, but much closer to the 'sun' pyramid. We interpret the first four are: Jupiter, Saturn Uranus and Neptune; the outer planets. The other three are closer to the *sun;* they are Mercury, Venus and Mars; the inner planets. The 'Earth' and Moon are clearly represented, as explained. In addition, there are four other small pyramids on either side of the Temple. These, we think represent the two equinoxes and the two solstices and the four seasons resulting from the astronomical arrangements the three objects can have in space.

Legend and folklore assign the feathered Serpent Quetzalcoatl, God like attributes similar to those given to the sun in other cultures. In Aztec culture, the feathers are associated with air and wind, but also as the reincarnation of several sun Gods [19]. As explained early on, the Temple to the God Quetzalcoatl, at Chichén Itzá, is highlighted by the sun during the equinox; the sun casts a progressive shadow from the top of the pyramid down to the ground. The shadow appears shaped like a serpent due to the pyramid terraces' which cast the shadow onto the side of the pyramid's northern steps. At the bottom of the steps is the head of the God Quetzalcoatl. The Temple celebrates a solar phenomena descending from above: The God Quetzalcoatl is the sun! The *traditional* pyramid of the Sun is the Earth receiving it and everything which is derived from it, including the phenomena described in this article.

The Aztecs or their ancestors may have witnessed the design and building of these monuments. Now with almost certainty we can say these engineers had flight capability. The ancestors to the Aztecs witnessed the builder's flight landings, which may have prompted the name "The Place Where Men Become Gods" Teotihuacán in Nahuatl.

Discussion

The results presented stand out, not only against everything that has been written about Teotihuacán, but also what is *commonly* known about our history. The sun positions calculated in antiquity, at first, were suspect which would result in flawed age data. The sun moon calculator we were aware had never been tested for dates prior to the twentieth century. However, our tests around the resulting age date confirmed the uniqueness of the sun/moon phenomena encountered. Also, the geometry of the celestial objects' positions in space matched the layout's physical geometry on the year the method indicated it was designed; which could not happen by chance. The date the eclipse phenomena occurred in was tested against the *Eclipse Season Cannon* [11]. The Saros cycle in the Cannon was calculated backwards in time; the eclipse phenomena coincided with the published data. The data presented thus far does not stand alone; the work at major sites, show similar results at El Infiernito, Colombia, Mohenjo Daro, Pakistan, Machu Picchu, Perú, The Pyramids of Giza, Egypt, The Temple of Kalasasaya, Bolivia, Stonehenge, UK and Tifariti, Western Sahara. All of which yielded similar data with eclipses that align with the monuments on their design dates or contemporaneously to the result dates.

Significantly, all these sites also reflected earlier dates from those commonly assigned to them via other methods. There are various investigators, Artur Posnaski among them, who claim the age of various monuments is still in doubt, the argument still continues. The analysis of results obtained at the archaeological site El Infiernito in Villa de Leyva in Colombia is discussed in the following section.

El INFIERNITO OBSERVATORIO SOLAR, COLOMBIA
(SOLAR OBSERVATORY)

This site is a megalithic circular -possibly spiral- arrangement. The type of spiral, as depicted in Figure 10, is common among those found on the Nasca plain. The site has the peculiarity of being laid out with stone megaliths carved to have a phallic appearance. The largest one at the center measures about 15feet tall. Remains found around them were C_{14} dated at about 2500 - 2880BCE [15]. We have found the date of the site's design, via the methodology to be between 8460 BCE and 7495 BCE, a time span of about 52 lunar saros cycles. The sun and moon alignments have yielded a result that would suggest a change in the site's name; from a solar observatory to a solar and lunar observatory - it actually records their positions in ancient times. This distinction is important and it will be demonstrated; we will show the sun and the moon were accurately tracked those years and the year 8082BCE which was also selected for analysis. On that year the sun and moon, both, reached maximum latitudes simultaneously. During the summer solstice on Sep 14 the sun reached 24.658°N and the moon latitude 29.718°N a *major lunar standstill*. The standstill alignment was marked at the site with menhirs: M-26, M-27, M-7, M-9, the Dolmen, M-36, M-22 and M-23 (our nomenclature, absent any other). The direction of the standstill is indicated with the line to the NE with azimuth of 12.5°. The first *major standstill* that year happened twelve hours before the vernal equinox on June 12 at latitude 29.8040°S, it was marked with the dolmen and menhir M-15; the line pointing SE. These are shown in Figure13.

Incidentally, in the study we also came about *an explanation* for a local legend. NE from the site, high in the mountain at 12,500 feet above sea level is Lake Iguáque. This lake is where the legend tells the origin of the Muisca peoples took place: after the moon shone brightly over its waters, the Goddess Bachué emerged from the lake holding an infant whom she later married. In the analysis we show how the Full Moon, on December 1, six days before the autumnal equinox rose precisely over the lake and its position was recorded with megaliths M24 and M13 at El Infiernito in the valley below.

Figure10 -El Infiernito's menhirs form one of several possible spiral arrangements

The site has traditionally been described as a solar observatory for the obvious east west alignment of two rows of menhirs which appear to align with the sun during the equinox; shown, in Figure 17. A poetic explanation of the site by anthropologist Eliécer Silva Celis appears on a billboard next to the central megalith:

OBSERVATORIO SOLAR
"The rectangle included between the two rows of columns placed in an east west orientation, probably, constituted a sacred site destined to receive the Sun as it appeared in the horizon and to follow it and observe its movement to its zenith, a phenomenon the Muiscas determined by means of pillars planted in open sky with the purpose marking with their shadows on the ground the solstices and equinoxes. These astronomical events could have been celebrated with spiritual rituals related to fertility and conception".
Translation and Photo Figure11 by the author 2016

As this study shows, the belief that the ancient designers of monuments like this one might have wanted to *capture* in their designs common place yearly astronomical phenomena is perhaps misguided. It is a questionable purpose for sites like this one and other designs, particularly those monuments which required *astronomical* amounts of resources and enterprise to execute. Capturing rare astronomical phenomena which reflects the positioning of the sun, moon and earth would, otherwise, justify the effort.

We interpret the megalithic phallic theme, which is found in many places around the earth, as a message which could mean: We the authors of these monuments are *human*, not unlike you, physically and biologically. Incidentally, there are also an innumerable number of monuments whose entrances are vaginal in shape; the entrances to the Pyramid of Khafre and to the Newgrange dolmen are prominent examples, here at El Infiernito the dolmen is below grade with its entrance facing south and is guarded by two rows of phallic megaliths. Later we'll see how this inner sanctum is illuminated by the moon at various times. Anthropomorphic shapes are, also, found throughout the Inca world in the shape of trapezoidal gates and windows and fountains, such as, The Princess' Bath in Ollantaytambo, not to mention the thousands of stele with explicit human sexual organs, both male and female. These are found at various archaeological sites throughout the Americas; in Colombia at San Agustín, the most prominent site with over 600 stele. There are also innumerable numbers of caves whose interiors are illuminated in obvious ways. And most prominently at El Infiernito (Little Hell) named by the Catholic Spanish Conquistadores perhaps horrified by the sight of over forty erect phalli. We believe the central megalith and the Dolmen could represent the Sun and the Moon; male and female respectively. Regardless of the interpretation one might want to give them, what is really remarkable about the site is the interplay of the sun-moon events recorded in the arrangement of the phallic menhirs. The sun and moon's position at those times is a measurable physical reality we will show is reflected in their arrangement. Among these records are; eclipses of the sun and moon also the New and Full Moons; found in the span between the years 8460 BCE to 7495 BCE, the details of which will be discussed. Explorer Alexander Von Humboldt, a Prussian geographer and naturalist, (1769-1859) was perhaps the first to suggest the menhirs locations reflected the alignments of the sun and the moon [21].

For the study of the Infiernito site we numbered the megaliths forming the *ring*, originally from one thru twenty four, not being aware of any other published nomenclature. On June 2016 we visited the site and confirmed the location of 24 megaliths and added seventeen more which were not clearly visible with Google©earth, in particular those which had fallen. During our visit we encountered some objection by the site's manager to our taking measurements. Apparently studies at the site need to be sanctioned by the Pedagógica University. In the end we were able to confirm the location of each megalith in the site using GPS.

The site consists of four distinct megalithic arrangements: the largest arrangement is comprised of at least forty one megaliths. There are several 'small' or buried ones which were not catalogued; these form the main *ring*. The central megalith whose importance had been confirmed in our earlier work; forms part of a global 'set' of megaliths carved with an explicit phallic motif. The largest one is located in China near Nigxia, exactly on the same meridian as the main one found here, the Mandane Mangi Tungi in India and the Tevaifaara in Tahiti, among several others.

A second lithic arrangement is inside the entrance to the dolmen. The inner chamber of the Dolmen is found at 30^0 and seventy five feet from the main megalith M-24. In it there are two parallel rows of four, two feet high columns, approximately three feet apart. The rows are aligned south at 180^0. The central megalith and Dolmen alignment form a $30^0/60^0/90^0$ triangle. The dolmen's pillars and entrance are shown in Figure 12.

Figure 11 - El Infiernito Megalithic Park. Photo by the author 2016

The third lithic arrangement is found about one hundred feet west of the main megalith. It consists of two parallel rows of roughly round columns; the northern most row has a one meter high megalith at either end with about 40 smaller ones about one foot high or less in between them. The central row has 49-51 megaliths between one and three feet high, about half of them appear to be broken, however, there may be a rationale for the size difference; there are not enough pieces strewn around to account for breakage. Mr. Silva Celis who took part in the excavations in 1981 relates there are about 54 or 55 pillars in each row. He also makes a distinction between the two rows and indicates the center one appears to be much older; "two architectural periods"[20]. This observation appears to be supported by our data. The menhir rows are shown in the panoramic photograph, Figure17. The two rows are arranged precisely at 90^0 east-west. The fourth group, immediately south of the 'central' row of columns consists of six large *uncut* megaliths and several others scattered around which may have formed other east-west rows. Some of the larger ones were identified and labeled, as well. Mr. Silva Celis provides further detail of what he construes the purpose of the site might have been.

"7. By means of the columns' alignments and monoliths with fixed natural points, such as the Iguáque Lake... the chibcha priests (Muiscas) surely calculated astronomical events and phenomena, including eclipses." [20]. Translation by the author

In our previous work, cited earlier, we had determined that this site's location is one point on a great circle generated by the location and azimuth (51.89°) of the Ahu Tu'u Tahi in Easter Island. This great circle includes thirteen archaeological sites around the world; several within Colombia, the menhir at Fohet in France and not coincidental as will be shown later, the ancient city of Mehrgarh in Pakistan; which is also found in Teotihuacán's great circle discussed earlier. At the El Infiernito this great circle aligns with the central megalith M24 and megalith M10. The El Infiernito section of the Tu'u Tahi great circle is shown in Figure 15 as well as El Infienito's location in the hemisphere. The antipode point to El Infiernito is in the Indonesian islands of Kepulauan Seribu. The great circle follows the Island of Sumatra lengthwise, passing by the Aur Duri megalithic site and in the Island of Java the circle passes through the city of Jakarta.

Figure 12 -El Infiernito Dolmen Entrance Photo by the author 2016

EL INFIRNITO'S AGE

As with the previous site we initiated the study by generating the *super-analemma* for the site centered on the location of the main megalith. A section of the analemma curve relevant to the study is shown in Figure 19. To find the age of the site we selected the intersection point of the analemma with the great circle in the south west direction, which corresponds to the position of the sun at noon on June 21, 8460BCE. The other possible alignments for this site are those with the cardinal points which are explicitly marked; the EW with the menhir rows and the NS with menhirs M-24 and M-25. The M-25 menhir has fallen down, it was estimated its size is somewhat smaller than M-24, the central megalith. The north direction we judged to be at its footing on the opposite side of its phallic carving. The line extending to the north at the intersection with the super analemma occurs in modern times, to the east occurs in the future and to the south at over 12,500 years in the past; this will be analyzed in the future. The remaining alignment to the west provided the date 7495BCE. Figure 13 illustrates the intersection points of El Infierno's megaliths with the super-analemma: the year 7495BCE to the west and the year 8460BCE with the Ahu Tu'u Tahi Great Circle running SW to NE.

YEAR 8460BCE

The year 8460BCE was a showcase of astronomical phenomena; two lunar and two solar eclipses took place. Multiple eclipses during major or minor lunar standstills are common. A partial solar eclipse took place on March 31st, two weeks after the culmination of the winter solstice on March 12th.. The sun and moon travelled together near the Tropic of Capricorn, the maximum solar eclipse occurred when they achieved latitudes 23.57^0S and 23.64^0S respectively. The plot of the sun and moon positions is centered on the central megalith M-24. The eclipse was tracked from its start at sunrise 6:12:00, with the moon rising earlier about two degrees above the sun at virtually the same azimuth. At first contact at 8:00:00 the partial eclipse started, their alignment is marked with menhirs M-24 and M-18. The maximum occultation of magnitude .32 at 9:36:16 is aligned with menhirs M-24 and M-19 and the end of the partial eclipse at 11:20:56 is marked with the alignment of menhirs M-24 and M-20. Figures 14 and 15.

The following month on April 14 during the full moon a partial lunar eclipse took place, which started a half hour before the moon set just before sunrise, it was barely visible at this location. The sun, earth and moon aligned at 181.67^0 with an elevation difference at moonset of 0.38^0. The start of the partial eclipse across the site was marked when the moon was at about 292^0NW in alignment with menhirs M-01, M-24, M-18 and M-17 in the direction of the rising sun at about 110^0.

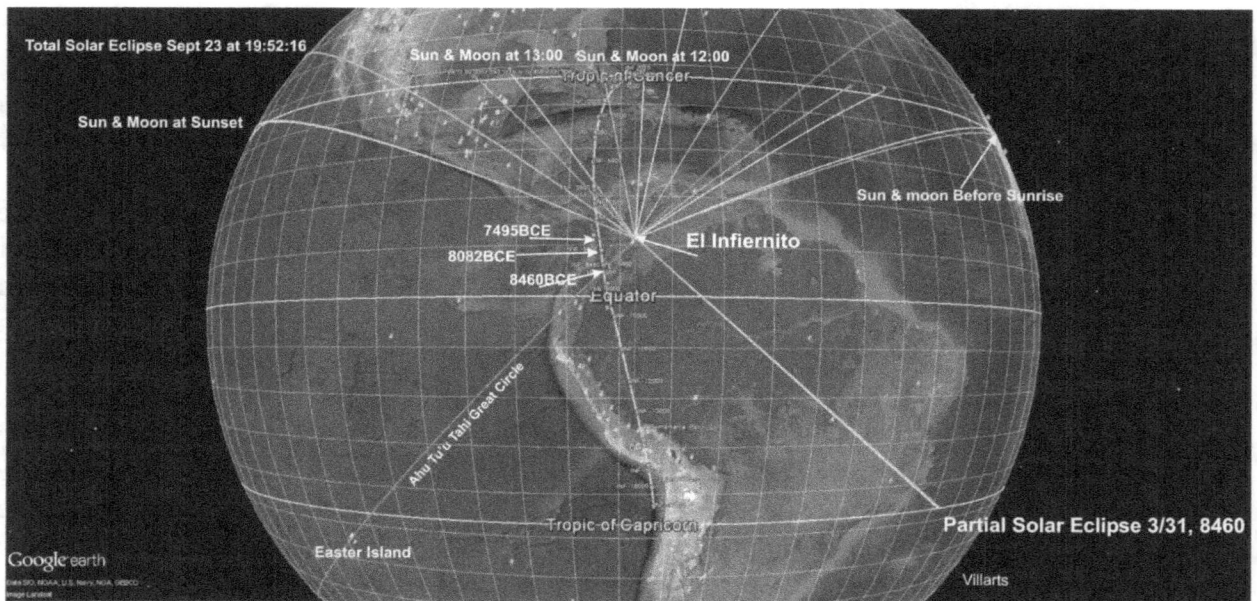

Figure 13. El Infiernito Great Circle and Super-analemma

On April 29th during the following new moon, the sun and moon aligned just before sunset but no eclipse took place. However the position of the New Moon was recorded. The sun and moon azimuths were marked with the uncut menhirs: the sun's with A1 and D2 to the SW at 246° and in line with menhir M-24, the moon's position with B1 and D2. The data appears in Table 4. This alignment repeated twenty saros cycles later, on June 4, 8082BCE; a total lunar eclipse took place and its phases were tracked with the central menhir and the central menhir row, as the data will show. This eclipse, we believe, determined the alignment of the central menhir row and the equidistance between the pillars. This is the *older* of the two rows as described by Silva Elis. The data and analysis for the eclipse are provided below under the heading; year 8082BCE.

The description of the site quoted at the beginning of this section, regarding the astronomical purpose of the site was based on empirical observation. Even in the year 8460BCE, at the right moment during the equinox the sun at one point visually aligns at 90° with these rows. The vernal equinox on June 15 of that year would appear to corroborate the common belief the two menhir rows aligned east west were set in this fashion to mark the equinox at sunrise. On that date in El Infiernito the sun crossed the Equator at 1:17hrs. with an azimuth of 76°, and altitude *below the horizon.* (Located in a valley shielded by a 12,500ft mountain- See Figure17). At sunrise it had a geometric elevation of -0.855° and its azimuth approximated the menhir row alignment with an azimuth of 89.84°. Correcting for the site's surrounding mountain's elevation, eight minutes later, the sun was at an altitude of 1.1° and reached an azimuth of 90.03°. Earlier we had argued that the work required to align menhirs to the equinox, a recurring annual event, would be superfluous exercise; which we now have established has been occurring for thousands of years. Instead, we will show the E/W alignment of the central row of menhir pillars, together with the central megalith, track a total solar eclipse in the year 8082BCE.

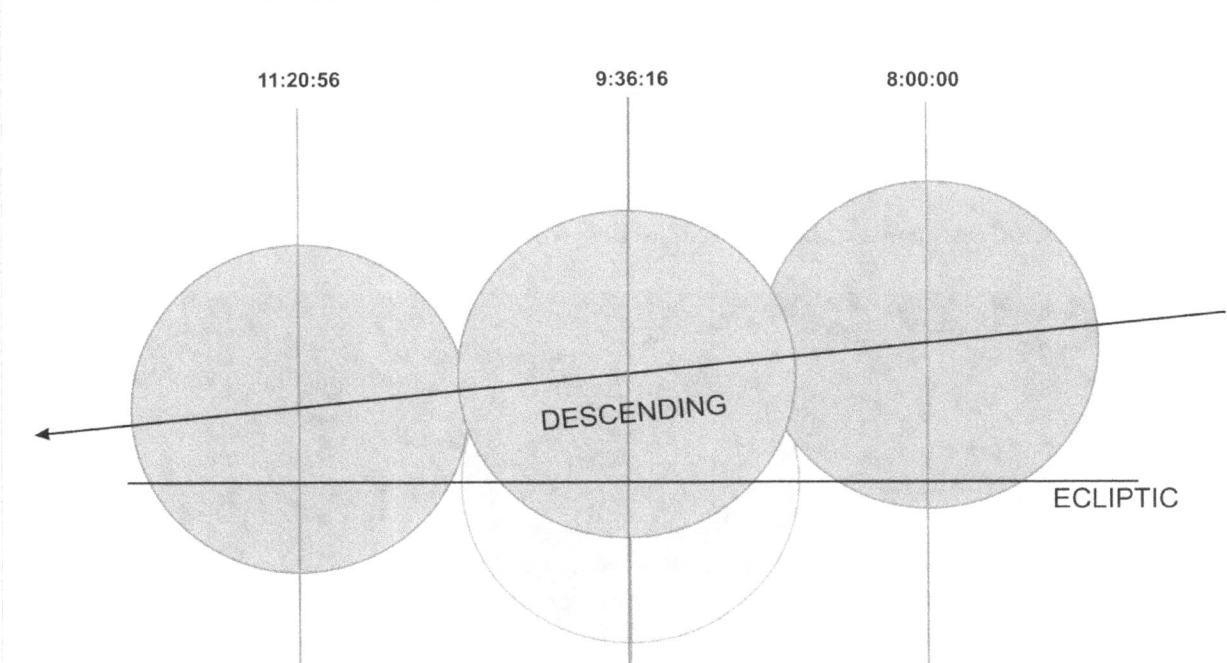

Figure 14 - Partial Solar Eclipse March 31, 8460BCE

A megalithic arrangement designed to track the phases of a total solar eclipse, a rare event, which in our estimation would justify the effort. The precision with which the menhir pillars were planted is demonstrated; it tracks the entire cycle of the motion of the moon and earth until the perfect alignment with the sun is achieved for a total lunar eclipse to take place. Before continuing the discussion of this remarkable event, we finish the year 8460BCE by mentioning the second solar eclipse which occurred after the summer solstice (September 12). The eclipse on September 23 was partial and only the close approximation and start were viewable at sunset from the El Infiernito when the sun was still 3.2^0 above the horizon. The sunset occurred eighteen minutes later at 18:18:12.

The maximum eclipse occurred over the Pacific Ocean at longitude 167.529^0E, then, the sun and moon had declinations of 24.257^0N and 23.973^0N respectively. The sun and moon positions are illustrated in Figures 15 and 16, lines pointing to the NW and the data appears in Table 4.

YEAR 8280 BCE

We make a parenthesis to introduce an astronomic event shared between El Infiernito and Mohenjo Daro. Our method reveals 8280BCE is the year Mohenjo Daro was designed; this monument will be discussed in detail in the next chapter. Reciprocal sharing of astronomical events between these sites repeats later on as will be discussed. On July 17 of 8280BCE a partial solar eclipse took place in Mohenjo Daro which had started late in the evening just before sundown. The eclipse could be seen at the time of its maximum occultation at El Infiernito; it was 11:09:11 at this site. See Table 5. At El Infiernito the position of the sun was *coincidentally* captured by the alignment of menhirs M-24 and M-10, the alignment of which is at the azimuth of the great circle which starts in Easter Island and connects El Infiernito with Mehrgarh located about 145 miles north of Mohenjo Daro. The moon's position was captured with menhirs M-24, M-34 and M-35. Surprisingly, 198 years later, about 11 saros cycles later, a total lunar eclipse took place on June 4, 8082, mentioned earlier, *due south of Mohenjo Daro*, it was recorded at El Infiernito but could not be seen there. It will be described next.

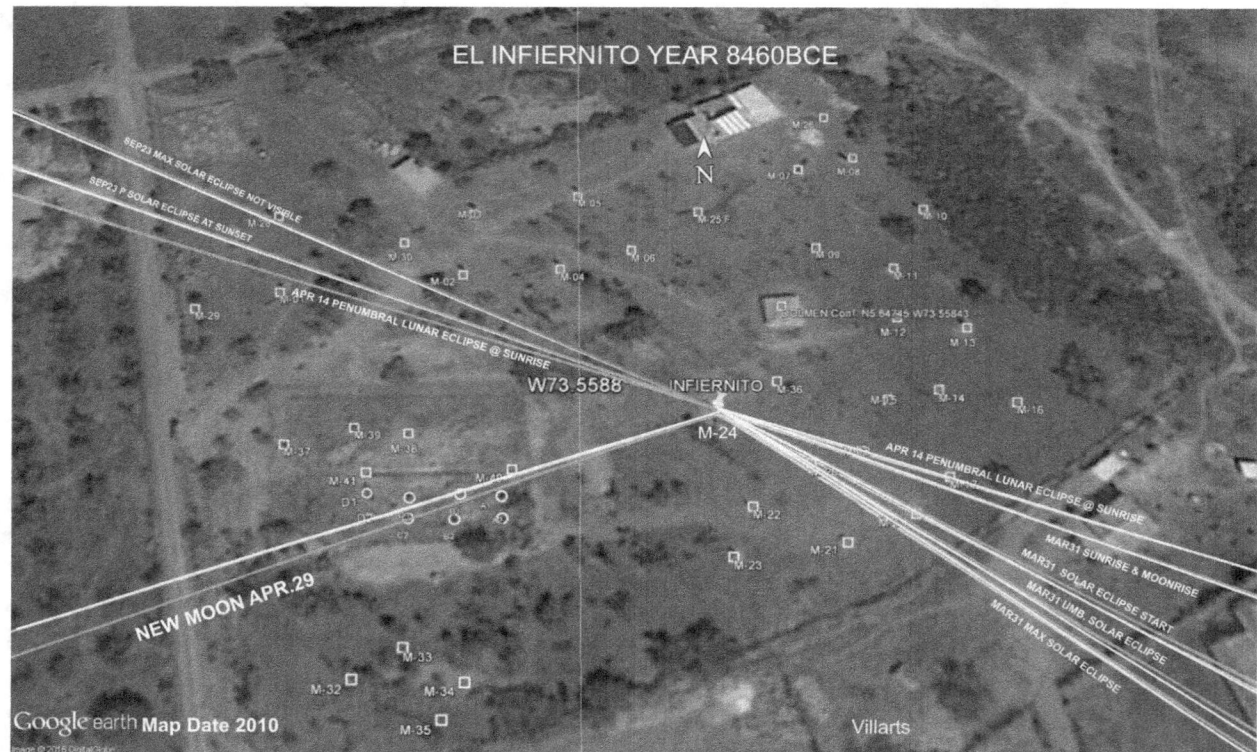

Figure 15 - Eclipses on March 31, April 29 and September 23, 8460 BCE

YEAR 8082 BCE

The year 8082BCE was arrived at by testing various years between the two dates indicated by the intersection points with the super analemma: 8460BCE and 7496BCE. Starting with the year 8460BCE the sun and moon positions were established according to the protocol and measured every 18.6 years; a saros cycle, starting from the total solar eclipse of 8460BCE. The year 8082BCE corresponds to the twentieth cycle.

A total lunar eclipse took place on June 4th. 8082BCE, its progress was accurately recorded with menhirs aligned to the positions of the sun and moon. As the eclipse progressed, at marked intervals each alignment was set using the central menhir as reference pivot point to record the sun and moon's azimuth in the sky. Other eclipses, solar and lunar events were also recorded in a similar manner, before and after this one. What makes this eclipse particularly significant is that *it could not be seen* at El Infiernito; an event remarkably similar to the one with the new moon at Mohenjo Daro and Teotihuacán. Mohenjo Daro, as described earlier, was associated with Teotihuacán in a similar manner, about one thousand years later; the New Moon in that case did not produce an eclipse, but the sun moon arrangement was cast in a magnificent design as was explained.

This time the maximum eclipse occurred minutes before midnight over the Indian Ocean 175⁰ due south of Mohenjo Daro. The ancient city of Mohenjo Daro is on a great circle which connects El Infiernito (M-24 and M-11) with Aur Duri in Indonesia. Aur Duri is an archaeological site with many megalithic pillars not very unlike the ones at Infiernito. Aur Duri is located in western central Sumatra, four kilometers north of the Equator, which makes it the archaeological site closest to the Equator. The importance this site has is discussed in detail in a previous writing [9].

GMT - 4:54:14   5.64726N 73.558585W

YEAR -8460	DATE	HOUR	AZIMUTH	S Long.	ELEVATION	RA	Declination	Longitude	HOUR	AZIMUTH	Ecl.Lat	ELEVATION	t-RA	Declination	Longitude	Ecl. Long	e.long dif	Decl. DIFF	Az. DIFF	Elv. Diff	RA D.ff
					SUN DATA								MOON DATA								
WINTER SOLSTICE	12-Mar	12:00:00	179.0368	269.734	59.548	269.707	-24.8	-73.021	12:00:00	66.1130	-4.7100	-43.3360	46.1200	13.0380	63.3920	46.7420	222.9920	37.8380	112.9238	102.8640	225.5870
SUNRISE	31-Mar	6:12:00	113.6182	287.468	-0.859	289.119	-23.585	14.902	6:12:00	113.5700	0.5070	0.1100	288.0810	-23.4400	13.8640	285.5900	1.8780	0.1450	0.0482	0.9630	1.0380
		7:00:00	115.1963	287.5	10.028	289.153	-23.581	2.904	7:00:00	115.2180	0.4640	10.5790	288.5570	-23.5010	-2.3070	286.0580	1.4420	0.0800	0.0217	0.5510	0.5960
PARTIAL START		8:00:00	118.7152	287.539	23.351	289.196	-23.576	-12.095	8:00:00	118.7360	0.4100	23.4400	289.0960	-23.5690	-12.1950	286.8440	0.8950	0.0070	0.0208	0.0890	0.1000
		9:00:00	124.6973	287.579	36.08	289.239	-23.57	-27.093	9:00:00	124.5800	0.3560	35.7800	289.5760	-23.6220	-26.7560	287.2300	0.3490	0.0520	0.1173	0.3000	0.3370
MAX SOLAR ECLIPSE		9:36:16	130.4985	287.604	43.637	289.266	-23.567	-36.659	9:36:16	130.1680	0.3220	43.1440	289.8560	-23.6450	-36.0690	287.6040	0.0000	0.0780	0.3305	0.4930	0.5900
		10:00:00	134.778	287.618	47.619	289.281	-23.565	-42.092	10:00:00	134.2570	0.3030	47.0450	290.0080	-23.6550	-41.3650	287.8160	0.1980	0.0900	0.5210	0.5740	0.7270
PARTIAL ENDS		11:20:58	159.3297	287.671	58.804	287.671	-23.557	-62.323	11:20:56	157.9680	0.2540	58.2770	290.5380	-23.1960	-23.1960	288.6080	0.9370	0.3610	1.8617	0.5270	2.8670
		12:00:00	177.2409	287.697	60.764	289.366	-23.554	-72.089	12:00:00	174.6130	0.1950	60.5680	290.7830	-23.5690	70.6720	288.9900	1.2930	0.0950	2.6279	0.1960	4.1170
	14-Apr	0:00:00	186.035	300.487	-74.361	302.945	-21.195	108.183	0:00:00	350.4000	0.4110	72.5290	118.0880	22.3580	76.6730	115.8160	184.6510	44.0530	164.3650	146.8900	184.8570
		5:30:00	110.5032	300.684	-10.636	303.171	-21.145	25.684	5:30:00	291.6930	0.6740	7.9690	120.2500	-22.2090	157.2370	118.6860	181.9980	43.3540	181.1898	18.6050	182.9210
PENUMB LUNAR ECLIPSE		6:00:00	110.9064	300.704	-3.653	303.192	-21.141	18.184	6:00:00	292.1120	0.6970	1.2710	120.5230	-22.1320	-164.4850	118.9460	181.7560	43.2735	181.2056	4.9240	182.6690
		6:10:00	111.1121	300.71	-1.331	303.199	-21.139	15.684	6:10:00	292.3210	0.7050	-0.9550	120.6170	22.1060	-166.8980	119.0330	181.6770	43.2450	181.2089	0.3760	182.5820
MID ECLIPSE		9:31:39	127.6231	300.848	44.985	303.342	-21.107	86.653	9:31:39	308.4740	0.8710	-45.0200	122.9600	21.5970	142.9640	120.8480	180.0000	42.7040	180.8509	90.0050	180.3820
	29-Apr	17:30:00	251.9429	315.394	6.836	318.16	-17.151	-154.605	17:30:00	249.7970	-2.1430	6.0990	317.7100	-19.5480	-155.0550	315.2830	0.1110	2.2170	2.1459	0.7370	0.4500
NEW MOON		17:45:00	252.4163	315.404	3.283	318.17	-17.128	-158.355	17:45:00	250.3040	-2.1560	2.7280	317.8600	-19.2910	-158.6650	315.4360	0.0320	2.1630	2.1123	0.5550	0.3100
		18:00:00	252.8179	315.414	-0.279	318.18	-17.125	-88.547	18:00:00	250.7580	-2.1690	-0.6490	318.0130	-19.2540	-162.2720	315.5900	0.1760	2.1090	2.0799	0.3700	0.1670
VERNAL EQUINOX	15-Jun	1:15:26	75.78059	0.000008	-67.694	0.000007	0.000003	85.218	1:15:26	157.445	4.973	77.064	276.179	-6.305	-68.603	206.329	206.329	6.305003	81.66441	144.758	206.179
		5:00:00	88.67272		-12.551	0.138	0.064	29.06	5:00:00	256.5		38.154	207.132	-6.844	-123.945			6.908	167.8273	50.705	206.994
SUNRISE		5:47:00	89.83778		-0.855	0.167	0.077	17.308	5:47:00	259.036	4.937	26.918	207.39	-6.995	-135.469	208.647	208.457	7.072	169.1982	27.773	207.223
		5:55:00	90.03239	0.19	1.136	0.172	0.08	15.308	5:55:00	259.382		24.999	207.437	-7.158	-137.428			7.238	169.3496	23.863	207.265
SUMMER SOLSTICE	12-Sep	5:43:30	64.97627	89.723	-0.839	89.695	24.8	19.99	5:43:30	239.27	-1.587	29.394	301.548	-23.252	-128.156	299.568	209.845	48.052	174.2937	30.233	211.893
PARTIAL SOLAR ECLIPSE	23-Sep	16:00:00	294.9203	101.4700	30.5680	102.6000	24.2730	-132.4930	16:00:00	294.5690	-0.2080	27.7200	99.4580	24.2340	-135.6360	99.4550	2.0150	0.0390	0.3513	2.8480	3.1420
Start of Partial Eclipse		17:00:00	293.6702	101.5140	16.4980	102.6480	24.2690	-147.9880	17:00:00	293.6190	-0.1550	14.5800	100.0003	24.2500	-150.6350	100.0190	1.4950	0.0190	0.0512	2.4140	2.6477
		18:00:00	294.0762	101.5560	3.2960	102.6940	24.2650	-162.4820	18:00:00	294.1420	-0.1050	1.6920	100.5700	24.2460	-164.6060	100.5820	0.9740	0.0190	0.0658	1.6040	2.1240
SUNSET		18:18:12	294.4827	101.5690	-0.8810	102.7080	24.2640	-167.0300	18:18:12	294.2470	-0.0900	-2.6090	100.7610	24.2470	-168.9770	100.7530	0.8160	0.0170	0.0803	1.7780	1.9470
		19:00:00	295.9411	101.5990	-10.2410	102.7400	24.2610	-177.4760	19:00:00	296.0180	-0.0540	-11.6360	101.2230	24.0650	-178.9930	101.1450	0.4540	0.1960	0.0769	1.3950	1.5170
MAXIMUM ECLIPSE		19:52:16	298.9556	101.6360	-21.7880	102.7810	24.2580	169.4620	19:52:16	298.9470	-0.0090	-22.6850	101.8440	23.9850	168.5250	101.5930	0.0000	0.2730	0.0680	0.8970	0.9370
Not viewable		20:00:00	299.5369	101.642	-23.467	102.787	24.257	167.329	20:00:00	299.506	-0.002	-24.289	101.94	23.978	166.682	101.708	0.0660	0.2840	0.0309	0.8220	0.8470
		20:30:00	302.1992	101.663	-29.874	102.811	24.255	160.032	20:30:00	302.048	0.023	-30.411	102.321	23.931	159.543	101.989	0.3260	0.3240	0.1512	0.5970	0.4900
		21:00:00	305.6407	101.684	-36.071	102.834	24.253	152.535	21:00:00	305.301	0.049	-36.354	102.715	23.891	152.416	102.27	0.5860	0.3620	0.3397	0.2060	0.1190
		12:00:00	156.4913	179.61	84.024	179.646	0.164	71.178	12:00:00	292.314	-4.708	-56.798	27.966	7.156	137.142	29.007	150.6030	6.9920	135.8222	140.8220	151.6800
AUTUMNAL EQUINOX	9-Dec	21:24:36	276.4139	179.9999	-48.485	179.9999	0.000001	147.64	21:24:36	81.496	-4.496	14.551	34.957	9.625	2.596	34.908	145.0919	9.6250	194.9179	63.0360	145.0429

EL INFIERNITO ECLIPSES YEAR 8460 BCE

Table 4. Eclipses on March 31, April14 and September 23, Year 8460 BCE

It is worth recalling the ancient city of Mehrgarh is on the great circle generated by Teotihuacán's Avenida de los Muertos, as is Mohenjo Daro; which is also on the Ahu Tu'u Tahi - Infiernito great circle- and the El Infiernito-Mohenjo Daro- Aur Duri great circle. There are three similar circles in all; all passing through Mohenjo Daro. See Figure 18.

On June 4th 8082BCE the sun and moon began to fall into alignment culminating in a lunar eclipse; at 4:00 hours they were aligned east west over El Infiernito, with the sun at 90.8⁰ and the moon at about 274.5⁰. Sixteen and a half hours later they had switched places; the sun was at 269.15⁰ and the moon at 89.995⁰. This peculiarity, we believe may have been registered with the first menhir row aligned east west, particularly in light of the fact than in the intervening 16.5 hours a total lunar eclipse took place on the other side of the world, 9:26 hours earlier from this location over the Indian Ocean south of Mohenjo Daro. The penumbral (P1) eclipse begun at 11:30:00 / 20:56:46, the greatest eclipse was reached at 14:20:00 / 23:47:41 and ended at 14:35:00 / 0:1:46, the umbral (U4) ended at 16:57:00 / 2:23:46 and the penumbral (P4) at 17:56:00 /3:23:06, at El Infiernito and in the Indian Ocean respectively, in Mohenjo Daro's local time. Figure 18 shows a global composite layout of the sun and moon's azimuths during the eclipse, from El Infiernito and from Mohenjo Daro.

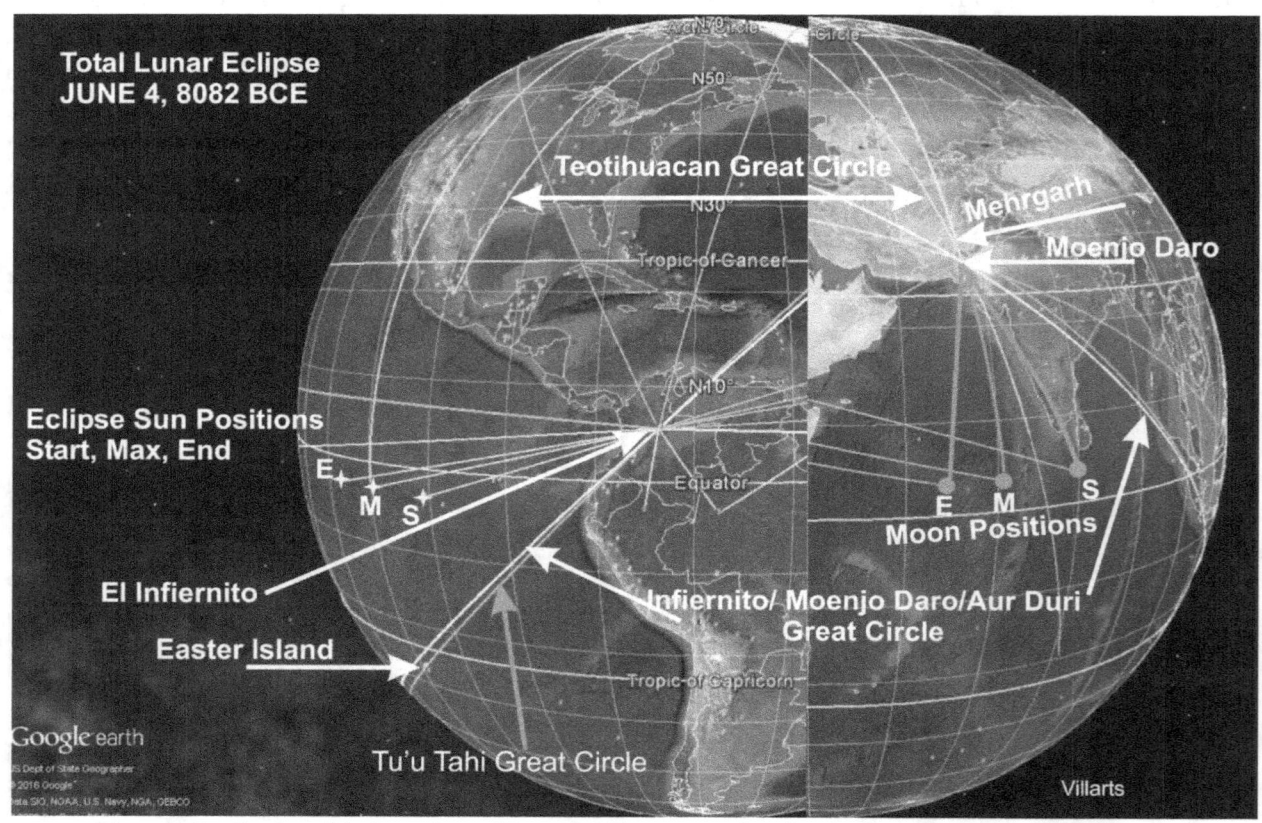

Figure 18– Geographic Location of Lunar eclipse at Mohenjo Daro, recorded at El Infiernito

Table 5 - El Infiernito & Mohenjo Daro Partial Solar Eclipse July 17, 8280BCE

The sun and moon positions at 4:00 and 19:45 hrs are shown in Figure23 each line is marked with the hour. The moon's position at 4:00hrs closely aligns from the west over the row of small pillars and the central menhir M-24, at the same time, the sun's position at about an hour before sunrise aligns from the east from M-16 to M-24. Nearly ten hours later the sun

and moon had moved westward while increasing the azimuth difference between them to the point where the earth's penumbra made first contact with the moon's circumference; their opposing positions getting closer to 180^0, diminishing from 185.347^0. At first penumbral contact, at El Infiernito 11:30 hrs, the ecliptic longitude difference was 181.477^0. The partial eclipse U1 ended at 12:52:00 hrs. the sun aligned with A2 and M-24. At 13:41:50 the sun from the west aligned with D2, C1, M-40, M-24 and M-13 and coincidentally with Lake Iguáque. As the eclipse progressed to maximum occultation at 14:20hrs (U2), the sun's *westward motion* causes its azimuth to sweep NW, thus aligning in succession with each pillar in the central row until the eclipse ends U4 at the last standing menhir M-41 at 14:46:30hrs. The moon's position at the eclipse's end is marked with menhir M-14. The sun's position at the end of the penumbral eclipse P4 at 16:57:00hrs. is marked with menhirs M-37 and M-38. The moon's position is marked with menhir M-15. The eclipse ended just before the moon rose over El Infiernito at 18:05hrs. After the eclipse, at 19:45hrs the sun and moon had switched positions; the sun was now below the horizon on the west at 269.46^0 and the moon had risen in the east at 90.37^0 at an altitude of 24.36^0. The data is presented in Table 6 and Figure 19.

4:32:32 4:54:14 5.64726   73.55859

9:26:46

inf/MD Time Dif	YEAR -8082 DATE	Moh Daro	SUN DATA HOUR	AZIMUTH	S.Longitud	ELEVATION	RA	DECLINATIO	longitud	MOON DATA HOUR	AZIMUTH	ECL LAT	ELEVATION	t-RA	t-DECLINAT	Longitude	Ecl Long	elongation	DEC Diff	Az Diff	Elv Diff	RA Diff
			17:56:16	255.5129	215.982	1.663	213.399	-14.24	180.395		36.166	5.225	-45.544	91.386	29.971	78.092	91.206	124.7760	44.2110	219.3449	47.2070	122.0130
MAJOR STANDSTILL	12-Jan	14:50:04	5:23:18	99.20718	205.442	-15.645	203.366	-10.362	31.357	5:23:18	336.887	5.113	63.021	86.561	30.194	85.448	87.189	118.2530	40.5560	237.6798	78.6660	116.8050
	4-Jun	13:26:46	4:00:00	90.7602	351.9090	-28.7660	352.6440	-3.3780	45.0340	4:00:00	274.5110	0.9300	23.6290	167.2680	6.3830	-140.3420	166.5620	185.3470	9.7610	183.7508	52.1950	185.3760
		13:56:46	4:30:00	91.4237	351.9290	-21.3000	352.6620	-3.3700	37.5320	4:30:00	274.8360	0.9050	16.3600	167.4700	6.2330	-147.6610	166.8400	185.0890	9.6030	183.4323	37.6600	185.1920
SUNRISE		15:19:12	5:52:26	93.2801	351.9840	-0.8440	352.7120	-3.3470	16.9590	5:52:26	276.2070	0.8360	-3.4810	168.1010	5.9210	-167.6540	167.6020	184.3820	9.2680	182.9269	2.6370	184.6110
NOT VISIBLE																						
Penumbral Eclipse Begins P1		20:56:46	11:30:00	145.6396	352.2110	79.2370	352.9190	-3.2530	-67.4990	11:30:00	333.6640	0.5520	-79.0200	171.7200	4.2020	111.3020	170.7340	181.4770	7.4550	188.0244	158.2570	181.1990
		21:26:46	12:00:00	189.2356	352.2310	80.9920	352.9380	-3.2440	-75.0010	12:00:00	13.3250	0.5270	-80.0210	172.0850	4.0860	104.1470	171.0130	181.2180	7.3100	175.9106	161.0130	180.8530
Partial Eclipse U1		22:18:46	12:52:00	238.7383	352.2660	73.0600	352.9690	-3.2300	-88.0050	12:52:00	57.4760	0.4830	-70.9140	172.7160	3.8310	91.7420	171.4570	180.7690	7.0610	181.2623	143.9740	180.2530
		23:03:46	13:37:00	351.5580	352.2970	62.8450	352.9970	-3.2170	-99.2580	13:37:00	70.5200	0.4450	-62.9480	173.2530	0.0000	80.9980	171.9170	180.3800	3.2170	181.0380	125.7930	179.7440
Iguaque Alignment		23:08:36	13:41:50	252.4728	352.3000	61.5790	353.0002	-3.2160	-100.5960	13:41:50	71.4850	0.4400	-61.7540	173.3160	3.6960	79.7200	171.9660	180.3340	6.8220	180.9878	123.3330	179.6942
		23:26:46	14:00:00	255.0035	352.3120	57.3630	353.0110	-3.2110	-104.1510	14:00:00	74.1760	0.4250	-57.7660	173.5220	3.5270	75.5010	172.1310	180.1810	6.7380	180.8275	115.1290	179.4890
Partial ends U1		23:41:46	14:15:00	256.7102	352.3220	53.7440	353.0200	-3.2070	-108.7610	14:15:00	76.0090	0.4120	-54.3330	173.6950	3.4600	71.9140	172.2710	180.0510	6.6670	180.7012	108.0770	179.3250
Greatest Eclipse & U2		23:47:41	14:20:55	257.2983	352.3260	52.5320	353.0230	-3.2050	-110.0110	14:20:55	76.6430	0.4080	-53.1820	173.7520	3.4380	70.7180	172.3260	180.0000	6.6430	180.6553	105.7140	179.2710
Mon @ So. Mohenjo Daro		23:51:46	14:25:00	257.6805	352.3290	51.3180	353.0260	-3.2040	-37.7030	14:25:00	77.0570	0.4040	-52.0290	173.8090	3.4160	69.5210	172.3640	179.9650	6.6200	180.6235	103.3470	179.2160
		23:56:46	14:30:00	258.1244	352.3320	50.1010	353.0290	-3.2020	-112.5120	14:30:00	77.5380	0.4000	-50.8720	173.8650	3.3940	68.3240	172.4110	179.9210	6.5960	180.5984	100.9730	179.1640
Total Eclipse Ends U3		0:01:46	14:35:00	258.5440	352.3360	48.8830	353.0320	-3.2010	-113.7620	14:35:00	77.9930	0.3950	-49.7130	173.9220	3.3720	67.1270	172.4570	179.8790	6.5730	180.5510	98.5960	179.1100
		0:13:16	14:46:30	259.4281	352.3430	46.0750	353.0390	-3.1980	-116.6380	14:40:00	79.9560	0.3850	-47.0380	174.0500	3.3210	64.3730	172.5650	179.7780	6.5190	180.4721	93.1130	178.9890
Partial Eclipse Ends U4		2:23:46	16:57:00	265.3249	352.4310	19.8730	353.1190	-3.1610	-149.2720	21:51:14	85.4460	0.2740	-16.1700	175.3860	2.7410	32.9950	173.7830	178.6480	5.9020	179.8389	36.0430	177.7330
Penumbra ends P4 SUNSET		3:23:06	17:56:20	266.9243	352.4710	-0.8570	353.1560	-3.1450	-164.1100	17:56:20	87.1900	0.2240	1.9420	175.9050	2.5710	18.6400	174.3380	178.1130	5.7160	179.6053	1.0850	177.2510
MOONRISE		3:26:46	18:00:00	267.0145	352.4730	-1.7680	353.1580	-1.7680	-165.0270	18:00:00	87.4240	0.2200	-1.0600	175.9350	2.4590	17.7510	174.3720	178.1010	4.2270	179.5905	0.7080	177.2230
	22:59:14		18:05:00	267.1363	352.4770	-3.0110	353.1610	-3.1420	-166.2770	18:05:00	87.5660	0.2160	0.1440	175.9760	2.4370	16.5380	174.4190	178.0580	5.5790	179.5703	3.1550	177.1850
MAJOR STANDSTILL	11-Jun	13:00:00		250.5874	359.058	72.289	359.144	-0.394	90.233	13:00:00	240.296	-5.284	-17.724	268.594	-29.982	179.216	269.726	89.3320	29.5880	10.2914	90.0130	90.5500
	12-Jun	0:00:00		273.2241	359.3430	-32.0610	359.4030	-0.2750	164.2370	0:00:00	136.7850	-5.2490	-55.3080	274.9840	-29.9370	79.8180	273.8860	85.4570	29.6620	136.4391	23.2470	84.4190
SUNRISE		5:47:41		90.0266	359.7400	-0.8390	359.7630	-0.1090	17.2740	5:47:41	162.2980	-5.1540	52.3570	281.3210	-29.5140	-61.1680	279.6700	80.0700	29.4050	72.2714	53.1960	78.4420
MOON IN DOLMEN		6:38:20		91.2770	359.7740	11.7660	359.7950	-0.0950	4.6080	6:38:20	179.9660	-5.1430	54.3230	281.6510	-29.4630	-73.5360	280.1680	79.8060	29.3680	88.6890	42.5570	78.1440
VERNAL EQUINOX		12:13:21		224.9563	359.9990	82.0450	359.9990	-0.00001	-79.1700	12:13:21	239.9690	-5.0630	-4.7460	284.2740	-29.2250	-154.8960	283.4580	76.5410	29.2250	15.0127	77.2990	75.7250
Start Partial Eclipse	18-Jun	11:38:00		140.6887	5.8230	85.8540	5.2910	2.4350	-70.9310	11:38:00	148.0160	0.7590	86.6650	4.4310	2.8170	71.7910	5.1950	0.6280	0.3820	7.3273	0.8110	0.8600
Maximum Eclipse		12:52:56		359.4143	5.8740	73.6180	5.3380	2.4560	-89.6690	12:52:56	262.2180	0.8210	73.1460	4.7280	3.1600	90.2790	5.8740	0.0000	0.7040	2.8037	0.4720	0.6100
		13:00:00		260.5798	5.8790	71.8870	5.3420	2.4580	-91.4360	13:00:00	263.1830	0.8260	71.4320	4.7410	3.1920	-92.0210	5.9380	0.0590	0.7340	2.6042	0.4550	0.5850
		13:30:00		263.8684	5.8990	64.4890	5.3600	2.4660	-98.9380	13:30:00	265.9710	0.8510	64.1210	4.8830	3.3270	-99.4160	6.2100	0.3110	0.8610	2.1026	0.3680	0.4770
End of Partial Eclipse		14:00:00		285.8231	5.9190	57.0540	5.3790	2.4750	-106.4400	14:00:00	267.6680	0.8750	56.7930	5.0160	3.4610	-106.8040	6.4810	0.5820	0.9860	1.8629	0.2610	0.3630

El Infiernito; Major Standstills January 12 and June 11, Eclipses June 4 and 18, 8082 BCE

Table 6. Major Standstills January 12 and June 11, Eclipses June 4 and 18, 8082 BCE

The sun's east /west alignment with menhirs M-24 and M-16 and the center magalith row at 90.0^0, we had argued is a yearly occurrence for the sun by itself, without particular merit. However, the simultaneous presence of the moon with the sun opposite each other at 180^0

causing a lunar eclipse and their alternating alignment east/west at 90⁰-270⁰, makes those alignments significant and worth recording. Also, is the fact the event occurred a week before the vernal equinox on June 12, six months after the major standstill of January 12th and a day before the second standstill in June 11.

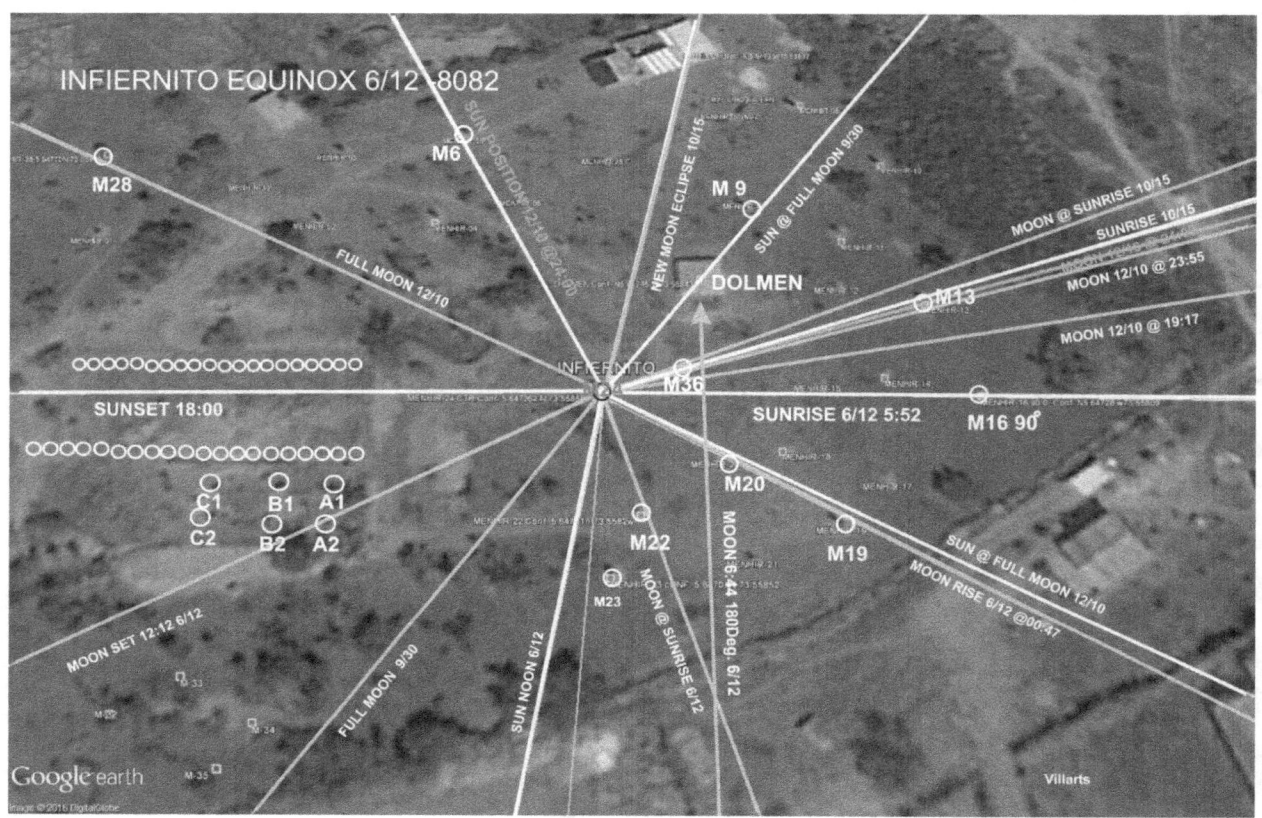

Figure 19 El Infiernito Lunar Eclipse of June 4th. 8082 BCE

This event was, worth marking with well defined menhir alignments; thus recording the event's importance, although, the eclipse could not be seen at El Infiernito. The implications these results have or that can be derived regarding the site's engineering, lead to some of the conclusions of this study. Following we'll see this alignment is reprised two weeks later by a partial solar eclipse.

The vernal (spring) equinox this year took place on June 12$^{th.}$ at 12:13:21; the sun had an azimuth of 224.95⁰ and an elevation of 82.05⁰. The previous day, the moon had reached a *major standstill* in the southern hemisphere with a declination 29.982⁰ S, had an altitude of -17.724⁰ and an azimuth of 240.3⁰. On the 12th the moon rose after sunrise (5:47hrs) and at 6:38:00 hrs it had an azimuth of 179.85⁰ and an elevation of 54.3⁰. This moon position is significant; since it was the last quarter and from this angle and elevation it illuminated the inner sanctum of the dolmen, also at that moment the sun was on the east at 91.27⁰ with an elevation of 11.7⁰. See

Table 6 and Figure 25. Six days later on June18th.the new moon resulted in a partial solar eclipse. The first contact occurred at 11:38:00hrs. it reached maximum obscuration at 12:52:56 hrs and last contact at 14:00:00. The partial solar eclipse ended when the sun and moon were both westward with azimuths of about 266^0 and 268^0respectively. The eclipse had magnitude about .15; it was not visible.

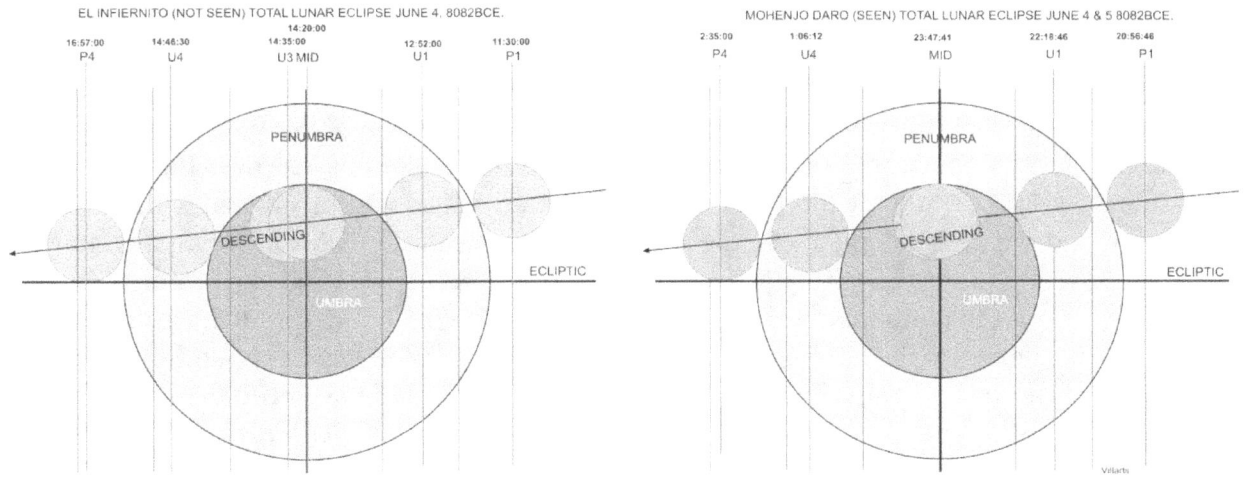

Figure 17 - Total lunar Eclipse seen at Mohenjo Daro recorded at El Infiernito, but not seen.

The culmination of the summer solstice occurred the midnight of September 14th at a solar declination of 24.658^0, and at the same time a *lunar standstill* at Latitude 29.718^0 took place; the moon's position was captured in an alignment of several menhirs with the dolmen, as was described earlier and shown in Figure 16. The following day on September 15th there was a New Moon at 16:37:46hrs, however the close approximation of about 5^0 did not yield a solar eclipse. Figure 24 shows the sun and moon's progression throughout the day, menhir M-12 marks the position of the sun and moon at sunrise. Menhir M-25 (fallen) towards the top marks north precisely; the graphic shows the positions of the sun and moon at noon time over it. The closest approximation of the sun and moon occurred at 16:37:46hrs when their ecliptic longitude difference is 0^0. The sun and moon positions at sunset are shown left towards the NW; their direction is marked between M-1 and M-28. The data is provided in Table 6 and in Figure 19. This graphic, less cluttered, shows, the now familiar Google overhead picture of the site with the menhir distribution and the dolmen -under a tiled roof. In the graphic, the line connecting M24 and M13 marks the direction of Lake Iguáque located in the mountains to the NE; see photograph Figure17. The direction of the Ahu Tu'u Tahi great circle is also shown with arrows pointing towards NE and SW. On September 29th.the moon was moving towards full moon, fifteen minutes before midnight it was located directly due south of El Infirnito at an azimuth of 180^0 and had an elevation of 55.3^0; from this position, once more, it illuminated the inside of the dolmen. The next day at 9:39:51 it achieved full phase when it aligned with the sun at 180.00^0, but it had set four hours earlier. However this alignment was captured by the alignment of

Menhir M-24, the dolmen and menhir M-9. This alignment is shown in Figure 20. There was no eclipse at this point, a perfect syzygy of the sun and moon did not occur; the moon was too high on its orbit.

EQUINOX

On the midnight on December 1st. the moon shone over El Infiernito with an azimuth of 77.87°, thirteen minutes later it reached an elevation of 60.123° and had moved northward reaching an azimuth of 71.477°, thus shining over Lake Iguáque. On subsequent days the sun and moon appeared together in the morning. Four days later, on December 5th., the moon reached standstill at latitude 29.545°. On December 7th. at sunrise the sun peeked over the mountain at an azimuth of 89.8°, while the moon was at 32.4°. Minutes later the sun aligned with the menhir row at 90° and the moon aligned with the dolmen and the central menhir M-24. In the afternoon the sun crossed the Equator at 12:58 hrs. Latitude 0.002°. As expected, at sunrise the sun was at nearly 90.0°, and the *alignment* was marked by menhirs M-24 and M-16. At sunset the sun was at 269.99° it aligned with the megalith rows from the west. See Figure 25 and Table 7. The confluence of the various astronomical events just described could have been the basis on which the Legend of Bachué was constructed: One can envision the activity which must have taken place for the erection of dozens of megaliths over a period of months, years, decades maybe even centuries; it could have caused a high level of consternation amongst the population. However, there is no record of their existence that early in history; the earliest known civilization is only about 4 or 5 thousand years old, according to the reported C_{14} dating.

4:54:14	5.64726	73.55859																					
YEAR				SUN DATA						HOUR				MOON DATA						D			
-8082	DATE	UT	HOUR	AZIMUTH	S.Longitude	ELEVATION	RA	DECLINAT	longitude	HOUR		AZIMUTH	ECL.LAT	ELEVATION	t-RA	t-DECLINAT	Longitude	Ecl. Long.	elongation	DEC. DIFF	Az. DIFF	Elv. Diff	RA Diff
SUMMER SOLSTICE	14-Sep		0:00:00	0.2105	94.8080	-59.6940	95.2920	24.6580	106.3250	0:00:00	Standstill	12.3480	4.9970	-54.2350	87.0450	29.7180	98.0780	87.2910	7.5170	5.0600	12.3375	5.4590	8.2470
MOON RISE	15-Sep		5:26:00	64.5374	95.0400	-5.2400	95.5470	24.6570	24.8570	5:26:00		60.1630	4.9640	-0.5630	91.0070	29.7150	20.3160	69.9800	5.0600	5.0660	4.3744	4.6770	4.5400
SUNRISE			5:45:31	65.1280	95.0540	-0.3470	95.5620	24.6490	19.9800	5:45:31		60.5330	4.9620	3.5120	91.1920	29.7130	15.8090	90.1400	4.9140	5.0640	4.5950	4.3590	4.3700
	NOON		12:00:00	339.8733	95.3210	71.0100	95.8560	24.6370	-73.6040	12:00:00		355.3050	4.9090	65.5740	93.6620	29.6170	-75.7980	93.2290	2.0920	4.9800	4.5683	5.4360	2.1940
			16:00:00	295.1711	95.4920	29.5390	96.0440	24.6800	-133.5800	16:00:00		300.6690	4.8670	27.9090	95.0730	29.6480	-134.5510	95.2080	2.0840	5.0180	5.4973	1.6300	0.9710
	NEW MOON		16:37:46	294.2431	95.5190	21.0030	96.0740	24.6290	-143.0180	16:37:46		299.4440	4.8600	20.0060	95.3590	29.5640	-143.7320	95.5190	0.0000	4.9350	5.2009	0.9970	0.7150
MOON IN DOLMEN	29-Sep		23:45:00	349.3705	110.2110	-60.6590	112.0670	23.1350	112.0760	23:45:00		180.1850	-4.6110	55.2670	286.3130	-79.0800	-73.6790	284.3890	174.1280	52.2200	169.1855	115.9630	174.2460
	30-Sep		1:29:46	34.1692	110.2860	-54.9170	112.1480	23.1230	85.8930	1:29:46		214.1730	-4.5680	47.5890	287.0700	-28.3320	-99.1850	285.4520	175.1660	51.4550	180.0038	102.5060	174.9220
MOON SET			5:47:56	66.4613	110.4690	-2.7580	112.3440	23.0940	21.8720	5:47:56		241.9350	-4.4550	-0.4240	289.4410	-27.9670	161.0310	288.1730	177.7040	51.0610	175.4737	3.3290	177.0970
FULL MOON			9:39:51	60.3918	110.6340	49.9270	112.5220	23.0680	-36.0880	9:39:51		234.7510	-4.3440	-48.3600	292.5960	-27.0760	143.9860	290.8340	180.0000	50.1440	174.3582	98.2870	180.0740
MOONRISE	13-Oct		5:56:00	69.8320		-2.1320	128.5460	19.8280	20.7660	5:56:00		65.7830		-0.5650	127.3720	34.1280	19.3920			4.0490	1.5670		1.1740
SUNRISE			6:01:29	69.9801		-0.6510	128.5500	19.8270	19.3950	6:01:29		65.9190		0.6390	127.4200	24.1140	18.2650			4.0611	1.4900		1.1300
NEW MOON			12:34:00	340.5848	126.1710	75.0280	128.8370	19.7540	-78.7160	12:34:00		347.3770	3.4270	76.2330	129.7430	23.3290	-77.8100	126.1710	0.0000	3.5750	6.4922	3.1770	0.9060
FULL MOON	29-Oct		21:04:50	294.4037	140.8730	-17.8630	143.5460	15.3200	154.6380	21:04:50		116.6230	-2.3220	35.5490	325.1580	-17.7980	-23.7500	320.8730	180.0000	33.1160	177.7007	73.4120	181.6120
PART SOLAR ECLIPSE	14-Nov		7:04:01	81.4925	156.5780	13.7840	158.5250	9.5800	3.3730	7:07:01		80.5660	0.9550	12.5660	159.8160	10.4070	-4.6640	156.3780	0.0000	0.8270	0.9365	1.2180	1.2910
			7:14:01	266.5062	156.5850	8.7030	185.6140	-2.5820	-154.5450	17:30:00		266.5910	0.9470	4.7140	181.8340	-2.9170	-158.5250	156.6640	0.0790	0.3350	0.0948	3.5890	3.9800
			8:19:01	82.2244	156.6270	31.0270	158.5710	9.5610	-14.1290	8:19:01		81.8020	0.9000	29.3710	160.2610	10.1690	12.4390	157.1840	0.5570	0.6020	0.7224	1.6360	1.6900
	30-Nov		24:00:00	341.8675	171.4450	-81.1780	174.0420	2.7400	109.1900	24:00:00		72.8730	3.3480	57.1090	23.6360	13.9930	-41.2260	26.5800	146.8650	11.2530	268.9945	138.2870	150.4060
MOON OVER IGUAQUE	1-Dec		0:13:00	1.5023	171.4540	-81.6010	174.0500	2.7360	105.9300	0:13:00		71.4770	3.3570	60.1230	23.6990	14.0530	-44.4220	26.7000	146.7540	11.3170	67.9747	141.7240	150.3510
LUNAR STANDSTILL	5-Dec		21:00:00	276.1085	178.3340	-42.2690	178.4870	0.5330	153.8200	21:00:00		47.2660	5.2220	-35.4480	87.6570	29.5450	62.3900	87.3320	91.0020	28.8480	228.8425	6.8210	90.8300
SUNRISE	7-Dec		6:06:00	89.80487	179.714	-0.757	179.74	0.12	17.214	6:06:00		32.461	4.755	61.616	105.938	29.061	-56.568	103.727	75.987	28.941	57.34387	62.373	73.802
			12:00:00	158.1438	179.9600	83.9370	179.9640	0.0170	-71.3050	12:00:00		298.1480	4.6290	20.4780	107.9560	28.2370	-143.8130	106.6430	79.3170	28.2200	140.0042	63.4590	72.0080
AUTUM EQUINOX	7-Dec		12:58:00	245.6174	180.0004	76.5290	180.0000	-0.0002	274.1920	12:58:00		297.6670	4.6070	8.0960	108.4280	28.1250	-157.3810	167.1210	72.8794	28.1252	52.0496	68.4330	71.5720
			13:30:00	255.0404	180.0230	69.0050	180.0210	-0.0090	93.8100	13:30:00		298.0270	4.5950	1.2800	108.7310	28.1000	-165.5170	107.3850	72.6390	28.1090	42.9866	67.7250	71.3080
SUNSET			18:14:16	269.99	180.219	-0.838	180.199	-0.092	-164.392	18:14:16		329.372	4.483	-51.452	111.938	27.173	127.348	109.711	70.5060	27.2650	59.3820	56.6140	68.2610

Table 7 – Moon in Dolmen Sep.29, Equinox alignment Dec.7, 8082BCE

The moon shinning over the lake is a cyclical occurrence, the fact that a major standstill happened that year, makes it rarer; the moon was higher in the sky than usual, which combined with the mystery of the megaliths' existence, their appearance or what they could mean, is the subject matter of how legends are created. The direction from which the moon shone then and the lake's location is prominently marked with two of the largest megaliths M-24 and M-13. The site could have served as the stage for rituals to the early settlers, as described by Silva Ellis. The early Muisca priests could have recognized some of the cyclical alignments of these magnificent phallic columns with the sun and moon deities.

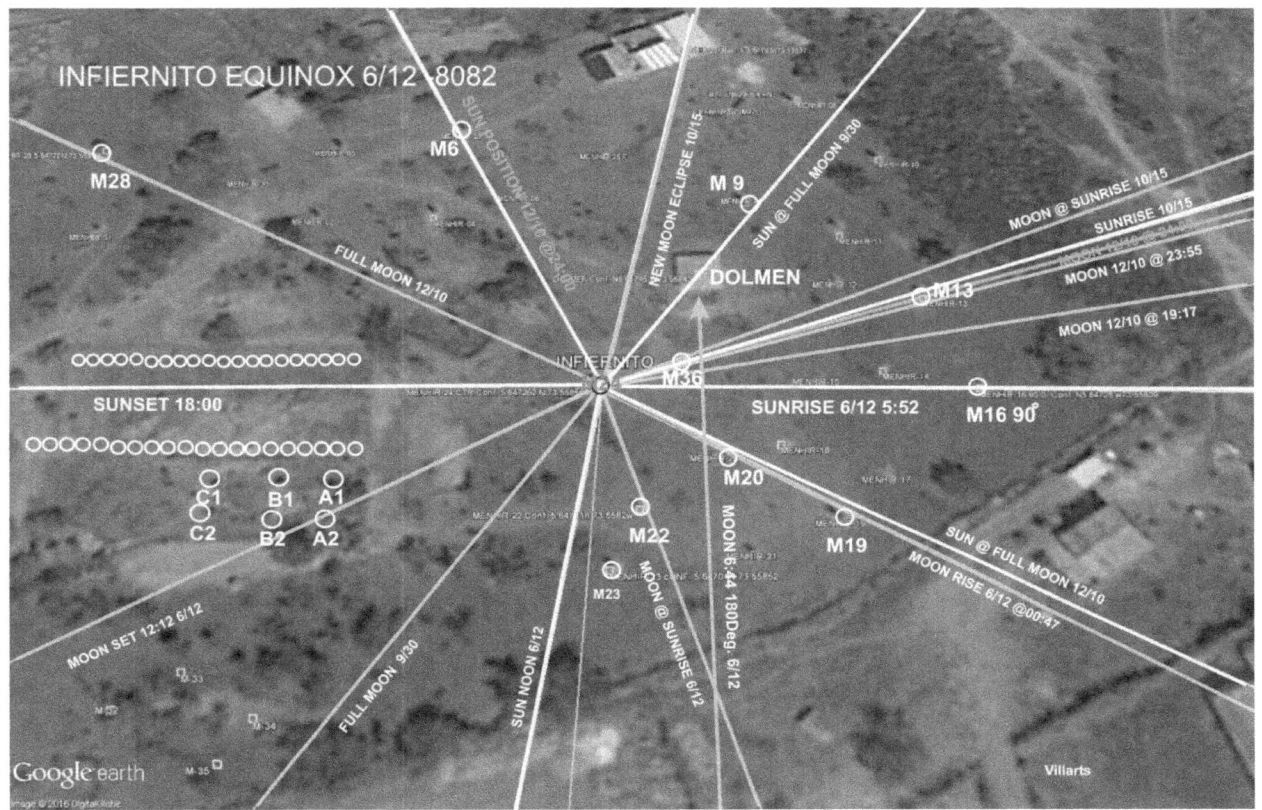

Figure 20 - Equinox and other phenomena Year 8082 BCE

YEAR 7495 BCE

The year 7495BCE was determined, via the method, by the east/west alignment of the menhir row with the sun, *at sunrise during the equi*nox and its alignment extended to the intersection with the super analemma. The equinox alignment is the commonly accepted empirical observation. In this section we test its validity, but, before we will look at two other phenomena.

In the year 7495BCE a partial solar eclipse of about magnitude .6 took place. Starting in November 3rd. at mid day; first contact was at 12:07:01 and had duration to last contact of approximately two hours, ending at 14:03:00. Sixteen days later after the solar eclipse, on the 19th. a partial umbral lunar eclipse took place starting before midnight on the 18th., first contact P1 was at 21:58:00. Nine minutes after midnight on the 19th. the moon achieved its maximum altitude of 76.85⁰ and had an azimuth of 180.1⁰.

Under those conditions, once more, the moon shone inside the dolmen, before the umbral eclipse started. Twenty five minutes after midnight it reached first umbral contact U1. The maximum eclipse was reached at 3:12:06 and the moon exited the umbra at 5:05:00. The last penumbral contact P4 was at 7:55:00. The part umbra eclipse lasted 4:40:00hrs and the total duration was 9:57:00hrs. These two events are depicted in Figure 21.

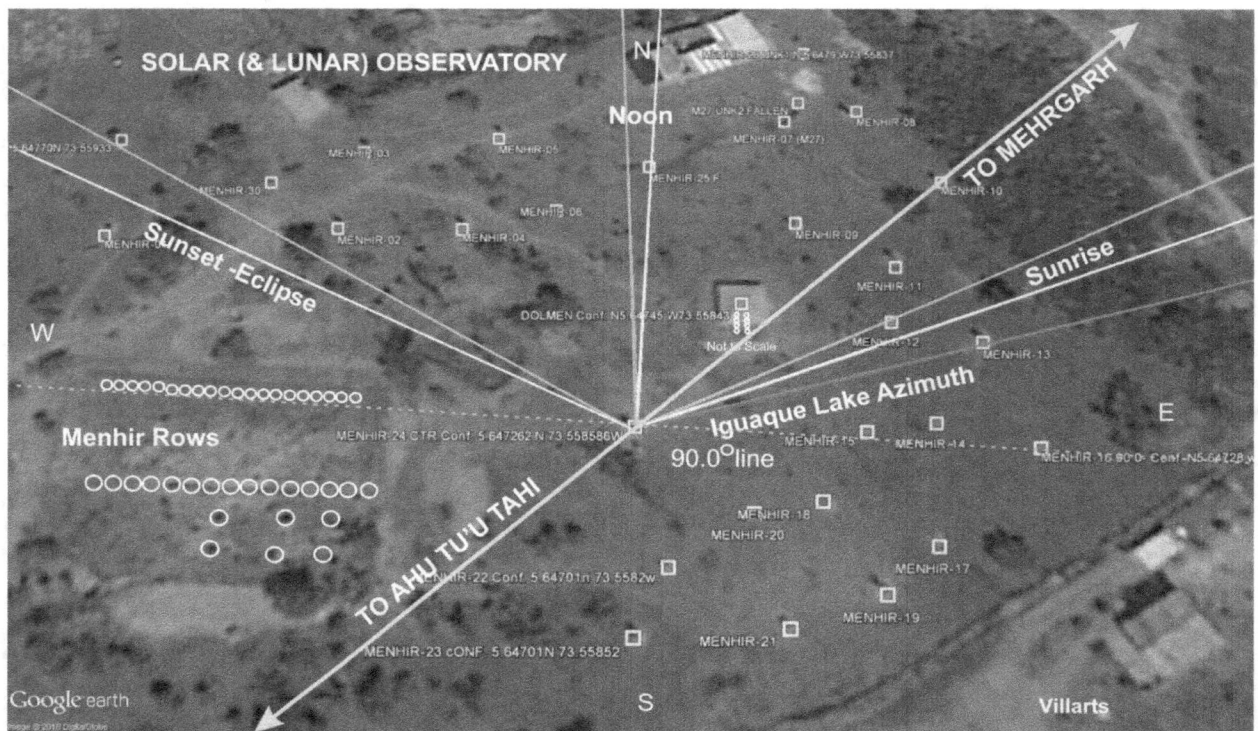

Figure 19. El Infiernito, Overview; arrow Great Circle, menhir alignments and Positions

EQUINOX ALIGNMENTS

The equinox is the point in time when the earth's position on its orbit is such that the direction of its axis tilt is tangential to its orbit. At this point the sun rays fall directly onto the equator at latitude 0.000^0; the sun's declination is 0^0. The beginning of the Vernal Equinox occurs the moment the direct sun rays move across the equator from south to north, as the earth continues about its orbit and its axis tilt is no longer tangential to it, the axis continues to point in a *fixed* direction to a point on the celestial sphere; currently 44' past Polaris. The direct sunrays change from a negative declination angle to 0^0 to positive declination. Likewise, the autumnal equinox occurs six months later when the sun's declination diminishes from its maximum latitude -the summer solstice- and crosses the equator from north to south, again reaching a temporary 0^0 declination before moving into negative degrees territory. The solar *longitude* at which these crossings -the equinoxes- occur *is not fixed*. The nodal points where it crosses are at the intersection of the equatorial plain with earth's orbital plain, the ecliptic; these nodal points are connected by an imaginary line which points to the celestial sphere. In 1AD the line through the ascending node pointed at the constellation Aries, so it was defined as the *Point of Aries*.

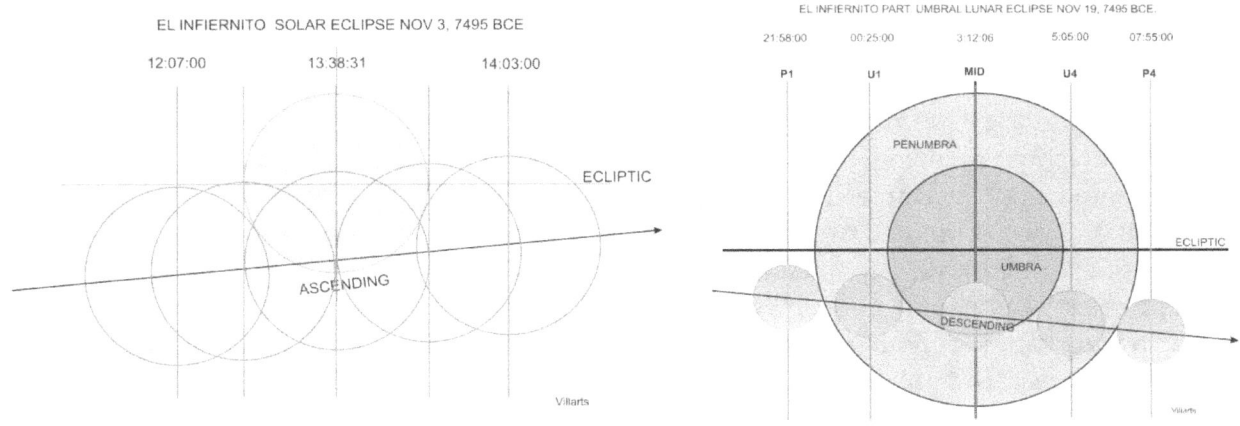

Figure 21- Partial solar Nov. 3rd and lunar Nov 19th, Eclipses

The point moves westward at about one degree every 72 years. The drift of the equinox from the *point of Aries* (currently Pisces) is *due to* the earth's axis precession with respect to the celestial sphere. Due to calendar corrections the equinox date was made static within one day. Thru millennia the crossing occurs at different hours of the day for any one reference location on earth year after year. Table 8a shows a sampling of equinox *equatorial* longitude crossing points in the recent era and data for about 9,000 years ago: 7207BCE. The table shows the hour the crossing took place and the solar azimuth at the time with respect to the main menhir M-24.

The alignment of the sun with a monument or menhirs during the equinox has changed through millennia, but, when these alignments are evaluated *today, within a few centuries, the month and day the alignment with the equinox occurs in appears to be static,* as shown in the table. The date was made static by convention. Table 8b shows the precise geometric solar crossing during the summer equinox 0.004^0 declination and 0.038^0 elevation, which aligned precisely at 90.0001^0 with the menhir rows. This occurred in 7496 BCE, the year before the ageing method had shown the alignment was designed 7495 BCE. Table 8b also shows similar data for the years of this study 8460, 8082 and 7495 BCE. The year 7496 BCE is highlighted in the table in lighter color. Data for the years 7493 thru 7480BCE is included to highlight that in the year 7487 BCE the equinox crossing occurred at longitude 15.385^0E, declination -0.066^0, *elevation 1.045^0*This elevation allowed the sun to peak over the 12,500ft mountain and align with the menhir row with an azimuth of 90.17^0; a visual alignment. El Infiernito is the archaeological site closest to the equator having menhir rows precisely aligned (90^0) east to west. The data presented confirms this section of the archaeological site was specifically designed to capture *the geometry* of the equinox.

Since the day Alexander Humboldt recognized the menhir rows at the site were designed to *align* with the sun at sunrise on the vernal equinox, naked-eye observations, empirically confirmed this, every year. However, mathematical analysis reveals this is only an approximation. The actual equinox, the crossover at 0^0 rarely happens at a point where the sun's equatorial longitude at crossover coincides with the correct azimuth for the sun to align at 90° with the site's menhir rows and the timing and altitude theoretically required for the sun's *visibility* at the site at sunrise, as described for the year 7487BCE. In El Infiernito the sun is not visible at the site until sufficient altitude is attained so the direct sun rays can clear the 12500ft. mountainous barrier in front of the site. Due to this, the data shows, the site designers must have aligned the menhirs geometrically, not visually on the day they laid out its design in the year 7496BCE.

We presented data for three important dates associated with El Infiernito's age starting on 8460BCE, the year 8082BCE twenty saros cycles later and the year 7495BCE. These years a good number of astronomical phenomena was found, in which the phallic pillars aligned to different types of celestial phenomena; eclipses, solstices and mayor and minor standstills. The data in Table 7b shows, that in none of those years the vernal equinox, at its actual solar declination of 0° latitude, coincided in alignment with the pillar rows at 90° azimuth; and neither in longitude (hour) with the *geometric* sunrise at the site. The first entry on the table is for the current year's vernal equinox in 2017. The equatorial crossover occurred on March 20 at 5:17:26 with the sun located at longitude 28.941°E and had an azimuth of 88.7503° with respect to the central phallic menhir M-24. Under these conditions the sun was -12.438° below the horizon, therefore it was not visible from the site. Less than one hour later at 6:10:00 the *sun moved* closer to El Infiernito and increased its declination from 0° to 0.014°. From this position it achieved an elevation of 0.643° a point at which it was nearly visible over the mountain and aligned at 90.05° with the menhir rows. The earth had rotated about 13°, close to one hour, after the equinox for it to align with the menhirs.

The 0:42:32 minutes it took for the sun to align may seem trivial, but it meant to us there would be a year, day, hour, minute and second when the sun's crossover point at 0° would align geometrically perfect with the sunrise and the menhir rows at El Infiernito. That date would be the day the rows were originally aligned: the site's age, or at least the menhir rows age, as was discussed.

Table 8b also shows the various dates we tested to find the 'perfect' fit. The geometry for the perfect alignment was previously discussed; the date when the sun at sunrise, elevation and longitude/hour, was June 7 at 5:51:50, 7496 BCE. On this date the sun aligned perfectly at 90°, with accuracy to four decimals, 12 minutes and fifty six seconds past the equinox with the sun at 0.004°N declination. At that point it had a geometric elevation 0.038° over the geometric horizon at El Infiernito. At this low solar elevation even atmospheric diffraction would not allow for direct line of sight of the sun over the 12,500 ft. barrier. This data supports the argument: the designers aligned the menhirs geometrically not visually.

YEAR	DATE	SUN DATA				
		HOUR		AZIMUTH	DECLINATION	longitude
	5-Jun	7:00:00		103.3662	-0.006	71.218
-7207	30-Nov	17:15:00		265.4894	0.000039	-175.667
1800	21-Mar	20:00:00		292.5935	0.002	119.931
	24-Sep	7:00		103.3155	0.003	71.274
1850	21-Mar	23:00		338.714	0.005	164.94
	24-Sep	10:00		145.5724	-0.006	26.279
1900	21-Mar	2:00		34.4456	0.008	150.052
	24-Sep	12:00		182.42	0.002	-3.712
1950	21-Mar	4:00		64.1567	-0.006	120.046
	23-Sep	14:00		218.6004	0.009	-33.703
1960	21-Mar	15:00		230.2565	0.008	-44.959
	23-Sep	1:00		21.27905	-0.006	161.296
2000	21-Mar	7:00		100.3079	-0.003	75.036
	23-Sep	17:00		259.6996	-0.001	-78.696

TABLES 8a - EQUINOX ALIGNMENTS & LONGITUDE

YEAR			SUN DATA					
			DATE	HOUR		AZIMUTH	ELEVATION	DECLINATI	longitude
2017	VERNAL EQUINOX		20-Mar	5:17:26		88.7503	-12.438	-0.000001	28.941
				6:10:00		90.04909	0.643	0.014	15.797
-8460	VERNAL EQUINOX		15-Jun	1:15:26		75.78059	-67.694	0.000003	-85.218
	AUTUMNAL EQUINOX		9-Dec	21:24:36		276.4139	-48.485	0.000001	147.64
	VERNAL EQUINOX		12-Jun	12:13:20		224.9400	82.0480	-0.00001	-79.1660
-8082	AUTUM EQUINOX		7-Dec	12:58:00		245.6174	76.5290	-0.0002	274.1920
-7497	VERNAL EQUINOX		6-Jun	23:50:00		357.8343	-84.349	0.00003	106.655
-7496	VERNAL EQUINOX		7-Jun	5:38:54		89.68508	-3.181	0.000001	19.638
-7496	VERNAL EQUINOX		7-Jun	5:51:50		90.00007	0.038	0.004	16.404
-7495	VERNAL EQUINOX		7-Jun	11:26:46		132.4161	81.66	0.000004	-67.411
	AUTUM EQUINOX		2-Dec	21:56:28		278.6393	-56.643	0.0000095	139.371
-7493	VERNAL EQUINOX		6-Jun	23:03:27		294.7287	-76.7	-0.000003	118.502
-7492	VERNAL EQUINOX		7-Jun	3:51:40		88.4889	-14.932	0.000004	31.448
-7491	VERNAL EQUINOX		7-Jun	10:39:50		106.8818	71.196	-0.000001	55.593
-7490	VERNAL EQUINOX		7-Jun			267.8453	20.818	-0.000007	-142.634
-7489	VERNAL EQUINOX		6-Jun	22:16:14		282.5393	-65.513	0.000004	130.308
-7488	VERNAL EQUINOX		7-Jun	4:04:29		87.15174	-26.68	0.000003	43.262
-7487	VERNAL EQUINOX		7-Jun	5:56		90.16996	1.045	-0.066	15.385
-7487	VERNAL EQUINOX		7-Jun	9:54		99.75901	59.742	0.0000025	43.783
-7486	VERNAL EQUINOX		7-Jun	15:40		266.3828	32.539	0.000015	-130.841
-7485	VERNAL EQUINOX		6-Jun	21:29:00		277.8045	-53.938	0.000009	142.118°
-7484	VERNAL EQUINOX			2:30:00		83.209	-50.09	0.00002	66.87
7482	VERNAL EQUINOX		7-Jun	14:53:32		264.4682	44.271	0.000002	119.015
-7481	VERNAL EQUINOX		6-Jun	20:49:44		275.1535	-42.251	0.000006	153.936
-7480	VERNAL EQUINOX		7-Jun	2:29:54		83.20384	-50.118	0.000001	66.895
-7480	VERNAL EQUINOX		7-Jun	5:52:00		89.95223	0.092	0.057	16.354

Table 8b Geometric Equinox Alignment on June 7, 7496 BCE

SUMMARY

Since the early days of archaeological study, it has been suggested some ancient monuments had as their purpose to celebrate some aspect of the cultures which flourished in the vicinity where the monuments are found. Absent any demonstrable reason for the existence for many of these sites, ethnocentric explanations abound. Those cultures may have adopted a site and explained their existence in their own terms. Whatever the purpose each site has been assigned, it has for the most part, been accepted with near devotion as it could be made to explain ageless traditions. Among those purposes from those which are attributed the local people's culture, the less mundane, are associated with astronomy. As such, the design of monuments are found to have elements that align with the sun during the solstices, equinoxes or the moon's standstills; which in turn are related to the seasons and weather. Therefore the most common conclusion reached by many is; they were built for agricultural purposes; as timekeepers of the growing seasons. This argument was and is still easy to believe, since those astronomical phenomena, could easily be argued affected the incumbent societies' daily lives; although, populations around the globe survived, even under the most arduous conditions, without having to build monuments of that nature. The early archeo-astronomers added to the accepted dogma by showing some degree of technical validity to those claims. Here we offer reasons for their existence which are more universal; more compelling, but, more important still is that the reasons offered are easily proven by any individual with basic skills on the subject.

 From the perspective we suggested in the prologue, the *Eagle's** point of view, without assigning a human culture rationale to their existence; we extended the *alignments concept* to include, not only their geometric designs as elements of those alignments but the structure's bearing and location on the earth's sphere: an all inclusive geometric *geocentric* analysis.

The monuments, we showed, not only were placed in locations on earth specifically chosen, but, their orientation angle(s) match the position of the sun, moon or both with respect to earth at an exact point in time. As a consequence, each monument froze in time the position each of the three celestial bodies had in space in the distant past, at the time each monument's *design* was laid out; from the these data we can tell the monuments' ages.

The method yielded important results at both sites. It shows their ages are much older than previously known; much earlier than any known civilizations had existed in the regions where they are found. The layout at both sites corresponds with astronomical phenomena which occurred in the years the method indicated they were designed. In addition, at Teotihuacán we found the entire site's alignment is to the new moon; the dead moon, therefore we propose a new name for the Avenida de los Muertos to become Avenida de la Luna Muerta. Also, the Avenue's length and azimuth connect the moon's phenomenon with it. Its relationship to the positions of the sun and Quetzalcoatl pyramids represents our solar system.

El Infiernito, a site with significantly simpler engineering the method shows it to be one thousand years older than Teotihuacán; however, the method used to track astronomical phenomena is similar. The site also contains representations of the sun and the moon. In the complete study web present an argument which relates these two monuments.

The study proved the basic hypotheses; the monuments are geodetic markers designed to record the positions of the sun, earth and moon at the time they were designed. Also their global positioning shows they were aligned geocentrically with the phenomena.

In the full version of the study (available at Amazon.com) we provide additional results of research conducted at Mohenjo Daro, Machu Picchu, Kalasasaya, Giza, Stonehenge and Tifariti which yielded similar results and provide support to our claim of a global design.

Conspicuously absent in this study are the results of the study of Göbekli Tepe, perhaps the oldest archaeological site on earth. The study is in progress; however its analysis is significantly more complex than any other site and for which many of its known megaliths have not yet been uncovered. From the 60 or so which have been cleared and restored we have tentatively calculated its age to range between 37,000 to 13,200 years old.

*The lunar landing module's name.

BIBLIOGRAPHY

1. Michell John; A little History of Astro-Archaeology. 1977 Thames & Hudson
2. Towers, Sherry; http://sherrytowers.com/2014/04/13/archeoastronomy-where-on-horizon-do-stars-sun-moon-rise-and-set/
3. Jens T. Satre; http://www.satellite-calculations.com/Satellite/suncalc.htm
4. Schlyter Paul; http://www.stjarnhimlen.se/comp/tutorial.html
5. Sir. Alfred Watkins; Early British Trackways, 1929 Amazon, Kindle
6. Villamarín Arturo; Nasca and Easter Island, An Ancient Global Plan Revealed. 2014, Amazon
7. Thom Alexander, Megalithic Lunar observatories, Clarendon Press, 1971 - Nature - 127 page 22
8. Worthey Guy; Tools of astronomy-precession. http://astro.wsu.edu/worthey/astro/html/class.html
9.   Villamarín Arturo; Timekeepers of Ancient Earth, p.152 - 2015, Amazon
10. https://en.wikipedia.org/wiki/Lunar_standstill
11. Espenak Fred; https://eclipse.gsfc.nasa.gov/SEhelp/moonorbit.html And; https://eclipse.gsfc.nasa.gov/LEsaros/LEperiodicity.html
12. Villamarín Arturo; http://www.earthsunexposure.com/INTIHUATANA.htm
13. Vsauce video: www.youtube.com/watch?v=IJhgZBn-LHg8
14. Gregory L. Possehl (11 November 2002). *The Indus Civilization: A Contemporary Perspective*. Rowman Altamira. p. 80. ISBN 9780759116429.
15. Eliécer Silva Celis; Investigaciones Arqueológicas En Villa De Leyva. Proyecto del parque arqueológico y botánico en villa de Leyva : sitio "El Infiernito" 1979
16. Jaime; http://www.aztec-history.com/about-aztec-history.html

www.ingramcontent.com/pod-product-compliance
Lightning Source LLC
Chambersburg PA
CBHW080135240526
45468CB00009BA/2456